ALEXANDER VON HUMBOLDT

Alexander von Humboldt

A CONCISE BIOGRAPHY

ANDREAS W. DAUM

TRANSLATED BY

ROBERT SAVAGE

PRINCETON UNIVERSITY PRESS

PRINCETON & OXFORD

Published by Princeton University Press
41 William Street, Princeton, New Jersey 08540
99 Banbury Road, Oxford OX2 6JX

press.princeton.edu

All Rights Reserved

ISBN 978-0-691-24736-6
ISBN (e-book) 978-0-691-24737-3

British Library Cataloging-in-Publication Data is available

Editorial: Eric Crahan and Rebecca Binnie
Production Editorial: Jenny Wolkowicki
Jacket design: Chris Ferrante
Production: Danielle Amatucci
Publicity: Alyssa Sanford and Carmen Jimenez

Jacket images: (Map): Atlas to Alex. v. Humboldt's *Cosmos*, 1851, edited by Traugott Bromme / Welcome Collection. (Portrait): Baron Auguste Gaspard Louis Desnoyers based on a drawing by Baron François Gérard, *Portrait of Alexander von Humboldt*, 1805. Courtesy of The Muriel and Philip Berman Gift, acquired from the John S. Phillips bequest of 1876 to the Pennsylvania Academy of the Fine Arts, with funds contributed by Muriel and Philip Berman, gifts (by exchange) of Lisa Norris Elkins, Bryant W. Langston, Samuel S. White 3rd and Vera White, with additional funds contributed by John Howard McFadden Jr., Thomas Skelton Harrison, and the Philip H. and A.S.W. Rosenbach Foundation, 1985, 1985-52-23859 / Philadelphia Museum of Art. (Mountains): *Mt. Chimborazo* and *Mt. Carguairazo*, drawn by Hildebrandt after a sketch by Humboldt, engraved by Edmund Lebel (1834–1908) © Royal Geographical Society / Bridgeman Images

This book has been composed in Arno

Printed in the United States of America

10 9 8 7 6 5 4 3 2 1

For Evis

CONTENTS

Revisiting Alexander von Humboldt

AS A GLOBETROTTING NATURALIST, multifaceted scientist, and international celebrity, Alexander von Humboldt (1769–1859) left a deep mark on his contemporaries and succeeding generations. They admired his passion for traveling extensively in Europe as well as to the Americas and Central Asia, crawling down mine shafts, climbing up volcanic mountains, and tirelessly investigating everything he encountered. The breadth of his interests, ranging from botany and geology to history and languages, has impressed people worldwide for two centuries. Humboldt's ideal of viewing nature as a 'whole,' interconnected with human society, has never lost its appeal, and it has drawn renewed attention in our time. For these and other reasons, which will surface in the succeeding chapters, Humboldt exerts an enduring fascination to this day.

This book provides a concise biography of Alexander von Humboldt and offers a compact guide to understanding his sprawling oeuvre. It introduces readers to the different stages of Humboldt's life and their historical circumstances, while illuminating the origins and trajectories of his scientific interests. The astonishing stream of publications that poured from his pen and their intentions deserve explaining, too, especially since Humboldt cultivated scholarly writing as a way of life.

Such a succinct overview already reveals some peculiarities of Humboldt's persona and thinking that may at first seem paradoxical. Many contemporaries regarded Humboldt as the most distinguished scholar of his time. But he never acquired an academic degree, taught at a university, or left us with a single game-changing discovery. The 'gentleman scientist' Humboldt made his most spectacular journeys under the protection of authoritarian regimes. Still, he fiercely protected his independence and became a critic of slavery and despotism. His approach to nature and society bore a Prussian and European imprint. Yet Humboldt repudiated nationalism and was influenced by non-European sources. He sent and received tens of thousands of letters and sought publicity like few scientists before him. Nonetheless, the ubiquitous savant closely guarded his privacy.

This biography expands on all these aspects while subjecting Humboldt to today's questions, inquiring into his scientific practices, social contexts, and networks of knowledge. It refrains from taking Humboldt out of his epoch and portraying him as a singular intellect way ahead of his time. Such heroizing tendencies have gained momentum in recent years, often catapulting Humboldt straight into the postmodern age. By contrast, my narrative does not ascribe achievements to this Prussian nobleman that cannot be substantiated or that he explicitly distanced himself from. Humboldt was not a revolutionary and pre-Darwinian advocate of evolutionary theory, nor did he 'invent nature.' The pendulum occasionally swings to the other extreme and reduces Humboldt to someone who saw non-European worlds primarily with 'imperial eyes.' Viewing him through a refined biographical lens avoids both mystification and vilification.[1] I want to suggest a more nuanced interpretation, portraying a multifaceted Humboldt whose thinking defies simple formulas.

First, Alexander von Humboldt did not fall out of the sky or emerge from a void. In every regard, he was a child of his times; the historical context matters. Humboldt's thinking was intertwined with practical and economic considerations, which audiences fascinated by his grand ideas have tended to underestimate. He also had to contend with political expectations. While extravagantly talented and attracting many admirers, he had many in-fluential predecessors and collaborators. His abilities needed a conducive environment and support from others to come to fruition. Humboldt operated within a wide range of social relations. Exchange with his brother Wilhelm was a constant in his biography. From this van-tage point, Humboldt becomes even more interesting historically. This also means taking seriously his Prussian and European experiences before sailing to America in 1799 and not reducing them to a prelude to his famous journey.

Second, Humboldt certainly emerges from the following chapters as an impressive polymath who embraced science and continuous research as his calling. Yet he was a complex personality. This biography focuses in a novel way on Humboldt as a human being, a mensch. That, too, matters for how we interpret his research and writing. More than just an intellectual, Humboldt was a sensitive and emotional man with a strong sensual streak, keenly aware of his own desires and frustrations. These human and emotional features directly influenced how he assembled and analyzed knowledge. Humboldt's science differed from the ideal type encapsulated in 'Humboldtian Science,' a term often used by historians of science.[2] His was a very personal practice of knowledge production, developing over time. For quite a while, Humboldt's science was rather experiential, marked by insecurities and subjective impulses that made him deviate from an ideal of objectivity.

Finally, Alexander von Humboldt was a figure representing an epoch—glorification and criticism, old and new, aside. His biography brings into focus some of the far-reaching transformations that unfolded over the course of his long life span of almost ninety years. Born in 1769, the same year as Napoleon, Humboldt lived from the eighteenth-century Enlightenment to the threshold of modern imperialism. He saw revolutions come and go in Europe and the Americas. Humboldt witnessed firsthand the ongoing enslavement of people but also the beginning of the end of slavery. He observed the collapse of the Spanish colonial empire, Napoleon's rise and fall, and the dawn of an industrial, technological age. A society that traveled by foot and on horseback, in coaches and on sailing ships, became in his lifetime one of railroads and steamships, both of which he came to use. By the end, Humboldt was as familiar with the revolutionary technology of telegraphic communication as he was with photography.

Humboldt recognized that trade, information, and human actions were becoming ever more closely interconnected across borders. At the same time, he experienced the emergence of nation-states and political ideologies that imposed new borderlines, both inside and out. These new divisions confirmed him in his determination to build bridges between cultures even though he insisted on maintaining his independence. The effect, ironic as it may appear, was that Humboldt became a privileged insider in the many spaces he traversed while remaining an outsider. Humboldt set out to compare nature and cultures globally. For that reason, too, he had to wrestle with a dynamic that he himself exemplified. Information had accumulated so rapidly, and become so specialized, that it was impossible to take in the many bits and pieces in a single glance. An increasingly self-assertive civil society demanded that knowledge be made accessible for all social strata and consumed in new formats. Humboldt's later years thus saw the emergence of a novel popular science.

Humboldt embraced all these challenges. At times, he was daunted by the sheer scale and breadth of what he had set out to accomplish. In this respect, too, Alexander von Humboldt belongs squarely in an era that lived through a "transformation of the world."[3] Like every individual from a bygone age, he retains an essential intransigence, notwithstanding his present-day appeal. I have tried to capture something of the flavor of his voice by quoting extensively from his letters, diaries, and publications. The chapter titles use the words of Humboldt himself, mentioned in the text.

This expanded English-language edition differs in several regards from the original one in German. A few details have been left out, and the introduction has been rewritten. Some sentences and passages have been added to clarify certain aspects. Two new maps and additional illustrations serve that purpose, too, and demonstrate the importance of Humboldt's travels within Europe. This edition also contains endnotes, primarily concerned with documenting direct quotes. With few exceptions, these are translated directly by me from the German or French to stay as close to the original meaning as possible. The endnotes therefore reference first the original sources; for ease of access, they are followed by available English-language translations, even if they differ slightly in wording. The selected bibliography at the end concentrates on publications in English. The new "Guide to Sources and Further Reading" gives readers some orientation in the ocean of materials anyone interested in Alexander von Humboldt will need to navigate.

1

Training the Mind

FROM CHILDHOOD TO
UNIVERSITY, 1769–1792

ALEXANDER VON HUMBOLDT was born on September 14, 1769, in the Kingdom of Prussia. Whether he entered the world in Berlin, as seems likely, or in Schloss Tegel, located to the northwest, remains unclear. Germany did not yet exist as a nation-state. In terms of population, Berlin lagged far behind other European cities such as Lisbon, Vienna, and St. Petersburg. It certainly could not bear comparison with London and Paris, the centers of the global powers England and France. But Berlin was catching up. The city's growth reflected Prussia's rise in political and economic standing in Central Europe. Frederick II ("the Great") had increased Prussian territory through his wars, even as he cultivated the arts and intellectual life in his summer residence of Potsdam outside Berlin. Humboldt's youth coincides with the last years of Frederick's reign and the decades following the Seven Years' War

(1756–63), which consolidated Prussia's position in Europe.

No one could have predicted that Humboldt would live to almost ninety. He far outstripped the average life expectancy in Europe, around thirty years for males in 1800.[1] If Alexander was a frail and sickly child, as the few sources indicate, he acquired remarkable physical stamina as he grew older. Humboldt would need it on the extended voyages that took him far beyond his native Prussia.

"Everyone is a product of their parents and the time," Humboldt wrote as a young man.[2] He was surely more than just that. His family background is nonetheless important for understanding his personal development and the independence he gained from an early age. Humboldt's mother, Maria Elisabeth, hailed from the bourgeois Colomb family. The Colombs had left France in the late seventeenth century as part of the Huguenot diaspora. They made their fortune in Brandenburg, the region surrounding Berlin, where the Huguenot refugees were granted religious tolerance by their fellow Protestants. They were useful for the ambitious Prussia due to their entrepreneurial initiative, specialist skills, and extensive trade connections. The Huguenots contributed to the policy of mercantilism, which the state pursued to promote the domestic economy, stimulate local industry, and improve the country's infrastructure.

Frederick II held the French language and culture in high regard. Humboldt took up both the mercantilist impulse and the cultural connection to France. He grew up speaking French as his mother tongue. For many years, beginning with his journey to South America, he conversed in French more than in German. For two decades, from 1807 to 1827, Paris was the epicenter of his life.

In 1760, Maria Elisabeth married the officer Friedrich Ernst von Holwede. He died only five years later. The inheritance substantially increased her wealth. She already owned a house left by her parents on Jägerstraße in Friedrichstadt, near the center of Berlin. This inner-city pad was to become Alexander's anchor in urban life. From her late husband, she inherited the Ringenwalde estate east of the river Oder, today in Polish Dyszno, and the small castle of Schloss Tegel, located around fourteen kilometers northwest of Berlin. Alexander divided his time between Tegel and Jägerstraße during the first two decades of his life.

In 1766, the widowed Maria Elisabeth embarked on a second marriage with the Prussian major Alexander Georg von Humboldt. His ancestors had distinguished themselves in state service without belonging to the old aristocracy. In 1738, his father petitioned the Prussian king for ennoblement. Alexander Georg himself spent several years at the Prussian court serving as chamberlain. Alexander von Humboldt would later intensify the connection to the royal house. Depending on the goals he was pursuing, he skillfully exploited both the aristocratic and middle-class sides of his background. Despite his occasional claims to the contrary, he was not a baron.

Maria Elisabeth gave birth to Wilhelm von Humboldt in June 1767. He too would live to a venerable age. Wilhelm died in April 1835 in Tegel at the age of sixty-seven. By that

FIGURE 1. Schloss Tegel (2015).
© Andreas W. Daum, Private Archives.

time, he was world-famous as a neohumanist philosopher, educational reformer, and linguist. When Alexander was born in 1769, the number of sons in the Humboldt family increased to three, for the Humboldt brothers grew up alongside a half brother. Their mother had brought Heinrich von Holwede with her from her first marriage. He was no match for Wilhelm and Alexander. What mattered more was that the brothers grew up in all-male company, even if they were occasionally joined by other students for lessons. A larger peer group of children and girls was lacking.

And the mother? The few surviving sources paint the portrait of an emotionally distant disciplinarian who kept her feelings close to her chest. They had "always been strangers" to each other, Alexander wrote when she died in 1796.[3] The contrast between Maria Elisabeth and her second husband, the father of Wilhelm and Alexander, seems all the starker and almost tragic in retrospect. The father was known to be jovial and entrepreneurial. Alexander was not yet ten when he passed away unexpectedly.

At first glance, then, Alexander's memories suggest a cloistered, love-starved childhood in which his "young soul" was "abused."[4] Yet we also hear of an Alexander who enjoyed dancing and drawing and occasionally toyed with the idea of becoming a soldier, a budding naturalist who delighted in the abundant flora and fauna he found while exploring the picturesque environs of Schloss Tegel. Undeniably, he was affected by a lack of emotional satisfaction and inner fulfillment, especially following his father's death. This situation drew him even closer to his brother. As they grew older, the brothers came to perceive ever more clearly both what bound them together and what set them apart. Wilhelm recognized that there was more to Alexander than just a fine mind. He understood Alexander's relentlessness and the questing ambition that drove him to ever greater challenges. For his part, Alexander admired his brother's intellect. Wilhelm remained for him a "splendid human being," albeit "too esoteric."[5]

As they grew into adults, both came to feel that they could be understood as polar opposites. On the one hand, Alexander was an attractive man of action with an unrivaled dedication to the natural sciences and geography, a human dynamo who refused to settle into the dull routine

of a career. On the other, Wilhelm, a gaunt and bookish
type, began to focus more on literature and philosophy.
He later took up service in the Prussian diplomatic corps
and in state administration. More attracted to men, Alexander never married or had children; fascinated by
women, Wilhelm embarked on a marriage with Caroline
von Dacheröden at the age of twenty-four and sired no
fewer than eight children, yet he remained a philanderer.

The convergences between them are no less important
than the differences. Alexander was interested in art,
history, and languages from an early age. In his later years,
he willingly served in the Prussian mining administration
and on state missions, in addition to fulfilling duties as a
member of the royal court. He communicated respectfully and often flirtatiously with educated women. For his
part, Wilhelm took a lively interest in his brother's research. In the 1790s, he and his brother even dissected
rabbits and rats together.[6] Wilhelm, who often subjected
himself to a ruthless self-discipline, later came to the
realization that his brother had outshone him from an
early age.[7]

Neither attended school. Live-in tutors oversaw their
education in all subjects. This was standard practice in
eighteenth-century aristocratic families and not uncom-
mon in affluent bourgeois households. Alexander and
Wilhelm benefited from their talented private tutors. They
were enmeshed in scholarly and literary networks that
expanded appreciably in the late eighteenth century. But
they intensified the atmosphere of male dominance in the
small Humboldt household and underscored its serious,

adult nature. This situation was not well suited for trialing new pedagogical ideas.

Johann Christian Kunth stood out among the boys' teachers. In his Lutheran background and appetite for *Bildung* (self-cultivation through education and knowledge), he embodied the central role that the Protestant bourgeois family played in the social basis of education and the state in Prussia. In the long term, Kunth had the greatest influence on the brothers. Maria Elisabeth hired him as tutor in 1777. She also entrusted him with the task of administering her fortune after her second husband died. Alexander benefited especially from his interest in geography. Kunth arranged for private sessions with Berlin scholars and opened doors for Alexander to new intellectual worlds. Christian Wilhelm von Dohm, later famous as an advocate for the emancipation of Jews, familiarized the brothers with questions of national economy, trade, and statistics. This thematic range allowed them to perceive global economic interconnections beyond Prussian mercantilism. Johann Jakob Engel introduced Wilhelm and Alexander to philosophy, which he taught them to understand as more than an inflexible system of propositions. It involved thinking in an ongoing dialogue with new ideas and grasping metaphysical categories as well as the mundane logic of cause and effect.

With their unbridled curiosity, the Humboldt brothers began to live out Immanuel Kant's programmatic definition of enlightenment as an intellectual process of empowering the individual, or, in Kant's words, "man's emergence from the state of self-imposed immaturity. . . . Have the courage

to use your own mind!"[8] Alexander was fifteen when Kant
wrote these epochal words in 1784. The chronic boredom
that Alexander is often said to have complained of, suffering in the confines of Schloss Tegel, was soon no more.
Only two years later, Alexander and Wilhelm began par-
ticipating in discussions of a reading group in Berlin, a
circle of like-minded intellectuals united by their commitment to enlightenment ideals. The brothers encountered
a Berlin that was emerging from its provincial slumber
and creating new forums for critical debate. The same
impulse was noticeable in many cities in Europe and the
United States, which had declared independence in 1776,
and even in the Spanish colonial empire.

Alexander's mental horizons were expanding along
with his opportunities for social interaction, not least
thanks to a remarkable trio of intellectually adventur-
ous, outgoing women and literary writers—Henriette
Herz, Dorothea Veit (later the partner of the philosopher
Friedrich Schlegel), and Rahel Levin, today better known
as Rahel Varnhagen.

The writer Gotthold Ephraim Lessing, the publisher
Friedrich Nicolai, and the Jewish philosopher Moses
Mendelssohn had spearheaded the so-called Berlin Enlightenment. In 1783 Nicolai had cofounded the Berlin
Mittwochsgesellschaft (Wednesday Society), an enlightened
discussion group. Mendelssohn, his extended family, and
his network of friends were important for Alexander's development. They familiarized him with the efforts of Jews
in Prussia to pursue the project of Haskalah: Jews should
be tolerated and treated as equals. Through the German

language and their cultural and economic engagement for the common good, they would find their place as an integral part of the German bourgeoisie. Humboldt lived and breathed this idea of tolerance, which was made easier for him by his own distance to faith, even in the undogmatic form to which Wilhelm remained open. Jewish families such as the Mendelssohns, Herzes, and Friedländers, to name but a few, enriched Alexander's hometown with their interconfessional salons. Humboldt's parents had valued the conversation cultivated there; now it was Alexander's turn to profit from it. He found the salon of Marcus Herz and his wife, Henriette, especially attractive. Marcus, a medical doctor and student of Kant, introduced him to the art of scientific experiment.

From October 1787, this intellectual ferment was supplemented by Humboldt's education at no fewer than four institutions of higher learning. Because there was still no university in Berlin, this involved moving further afield. Acceding to their mother's wish, the Humboldt brothers initially spent a semester at the old University Viadrina in Frankfurt an der Oder, east of Berlin. This stint was meant to prepare them for state service through the study of jurisprudence for Wilhelm and cameralism for Alexander. In the eighteenth century, cameralism denoted a comprehensive theory of public administration, including insights into economics, new technologies, and topics from the natural sciences. Although Alexander was unimpressed by his teachers, cameralist ideas played an important role in orienting his wide-ranging scientific curiosity to practical life and the social application of nature's bounty. The world of ideas merged with questions of the efficient use of knowledge.

While Wilhelm transferred to Göttingen University in the spring of 1788, Alexander returned to Berlin for a year of private study. He was in luck. His friendship with the slightly older, recently graduated physician Karl Ludwig Willdenow allowed him to deepen his interest in botany and steer it far beyond the borders of Prussia. Willdenow taught Alexander the importance of geographical and climatic differences between botanical zones, emphasizing the role of species migration for understanding plants in the entire ecosystem. Willdenow also familiarized Humboldt with the scientific results of transcontinental expeditions that had ventured to east Asia. Alexander was on the way to seeing natural phenomena in larger contexts. In hindsight, he emphasized the exchange with Willdenow as a decisive moment when he set his mind to traveling outside Europe and "nurtured immoderate desires for the faraway and unknown."[9]

Alexander finally left Prussia in April 1789 to follow his brother to Göttingen, in today's Lower Saxony. Although it had only opened its doors in 1737, the university there had soon become one of the most reputable in the German-speaking world. Alexander absorbed as much as he could: insights into mathematics and technology, archaeology, and the latest findings in the natural sciences. Their most important representatives in Göttingen combined empiricism with questions of natural philosophy. The physicist Georg Christoph Lichtenberg taught Alexander to pay scrupulous attention to detail with his measurements and tables in experimental physics. From Johann Friedrich Blumenbach, who lectured in medicine, Alexander picked up the idea of a developmental force or *Bildungstrieb*. Blumenbach took this to be a formative drive innate to every organism that led it to preserve itself, strike a balance between its anatomical structures and physiological processes, and reproduce. Both professors enjoyed a Europe-wide reputation. It was through them that Alexander came to know foreign students, although only men—women were not yet admitted to university.

Alexander was spreading his wings. Shortly before leaving for Göttingen, he wrote that he no longer felt "like a child in leading-strings." From now on, I will therefore refer to him mostly by surname. Humboldt resolved to become a "free being."[10] In the same year, 1789, many in neighboring France sought to realize this project on the political plane by turning against the late absolutist monarchy. On July 14, the storming of the Bastille kick-started the French Revolution. Of greater consequence than this symbolic act was the Declaration of the Rights of Man, announced by the French National Assembly at the end of August. It offered a heady mix of European Enlightenment ideas and the new principles of self-determination that others had articulated across the Atlantic in the American Revolution.

All men, the Declaration proclaimed in a phrase that heralded the dawn of a new era, were free, equal, and endowed with certain inalienable rights. But only men for the time being, and not even all of them. Many were still excluded from the society of free individuals equipped with equal rights in all realms of life. Among these were women

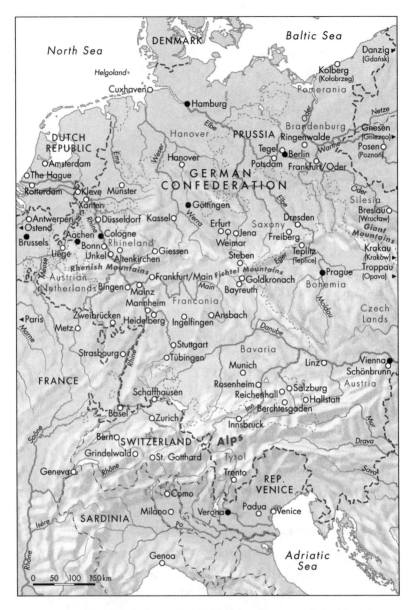

MAP 1. Locations Humboldt visited in Prussia and neighboring
territories, 1769–99 and 1804–59. By Peter Palm.

and the millions of enslaved people in the Atlantic world, as well as other underprivileged groups. In the following decades, Humboldt would come to understand both the appeal of the new, revolutionary ideas and their internal contradictions.

Against the backdrop of revolutionary turmoil in France, Humboldt set out from Göttingen in September 1789 on his first extended research trip, focusing on the Rhine River region.[11] He was accompanied by a Dutch student of medicine, Steven Jan van Geuns. Their route initially took them southwest to Heidelberg. They then passed through Mainz to the River Rhine, which they followed north past Düsseldorf, before finally returning to Göttingen via Münster. The tour established a template Humboldt would follow for decades to come. He would travel in the company of male friends handpicked for their scientific qualifications, journeying by coach and under sail, on horseback and on foot. He often spent the nights alongside his companions. From the 1790s onward, he packed instruments for measuring all kinds of physical phenomena. Touring Central Europe before the Napoleonic conquests and the downfall of the Holy Roman Empire in 1806, Humboldt encountered a patchwork quilt of different systems of government. At times, he crossed several political borders within a few days; he almost never had to show any papers. In Europe and America, nation-states and their borders only solidified over the course of the nineteenth century.

From 1789, Humboldt began complementing his existing interests with an impressive openness for the new and foreign. He sought and he found. On his journey with van Geuns, Humboldt had let no opportunity go to waste, from the quicksilver works in Zweibrücken in the Palatinate, west of the Rhine, to the Botanical Garden in Mannheim. Some items on his itinerary—the Rhenish basalt formations, for example—were the object of targeted investigation. Others, such as visits to old churches, arose more spontaneously. He arranged meetings with scholars through letters sent in advance. Many young noble and wealthier middle-class travelers did the same in the Age of Enlightenment, fired by their enthusiasm to uncover as many dimensions of past and present reality as possible. Humboldt pursued this goal with a particular zeal, fueled in later years by the ambition to document his travels more comprehensively than others and bring them more speedily to public attention.

Humboldt completed the manuscript of his first book publication only a few months after his journey with van Geuns. *Mineralogical Observations on Some Basalts at the Rhine* (1790) marked his precociously early transition from student to researcher. He showed off his philological acumen with lengthy excurses on ancient mineral nomenclature, even throwing in some musings on Rhenish viniculture. The twenty-year-old audaciously presented himself as an empiricist sweeping away the misunderstandings of older scholars. With his literary debut, above all, Humboldt became embroiled in one of the most important controversies in the late eighteenth-century natural sciences. Had the world of rock arisen through sedimentation of mineral deposits in the primordial oceans, as the so-called

FIGURE 2. Rhenish basalt formations near Unkel (2017).
© Andreas W. Daum, Private Archives.

Neptunists assumed? Or were the Plutonists and vulcanists right in arguing that rocks like basalt were volcanic in origin, the product of fire and movements beneath the earth's crust?

Having personally inspected the Rhenish basalts, Humboldt could now claim some authority in the vulcanism debate. The rock formations, water residues, and flora he found there spoke against the vulcanist theory. He may have had an inkling that the latter would prevail before long, for Humboldt tacked diplomatically between the two opposing interpretations. His comparative perspective— in this case of Ireland's Giant's Causeway and the Scottish

island of Staffa, known to him from his reading—always
gave him the option of revising his opinion and changing
his position. Throughout his life, Humboldt guarded himself in this way against scientific dogmatism.

After completing his studies in Göttingen, Humboldt
set out on his next journey in March 1790. This one lasted
longer, more than three months, and it took him further
away from Prussia—to England and revolutionary France.
Humboldt gained a prominent companion in Georg Forster. The two had met the previous year in Mainz. Forster
was a figure of some renown. He had accompanied his
father on James Cook's second circumnavigation of the
globe in the early 1770s. Many contemporaries, including
Humboldt, had enthusiastically read the written account
of his travels, *A Voyage Round the World* (1778–80). Forster would soon declare his support of the Jacobins, one
of the radical republican factions among the French
revolutionaries.

The revolution's impact could already be felt on the first
stage of their journey. This again led north from Mainz
through the Rhine valley, then on through the territories
of Jülich to Wallonian Liège, which had staged its own
revolution against princely rule in 1789. They stopped off
in Brussels, in the Austrian Netherlands; an independent
Belgium was still some forty years off. The northern Netherlands, their next destination, had long been a republic.
It was the eloquent and rather pretentious Forster, not
Humboldt, who documented their experiences in his
Views of the Lower Rhine (1791–94). The two men inspected factories that were integrated in global trade

flows, sourcing their materials from Spain and South America. They encountered the art of the Flemish Baroque painter Peter Paul Rubens and visited natural history collections. In Liège, they experienced firsthand the new mobilization of civic public opinion. Even "the common man politicized about the rights of humanity over his bottle of beer," Forster noted.[12]

Humboldt soaked all this up, although he never really warmed to Forster. He was repeatedly prone to bouts of melancholy. Yet the journey, he wrote about himself looking back, "trained my mind."[13] Later in life, Humboldt repeatedly emphasized how strongly Forster had influenced him. Forster set new standards for travel writing, combining vivid, subtle descriptions of geography and nature with sketches of the various ethnic groups he encountered. He characterized their customs, physical practices, and mythologies with great sensitivity. Forster and Humboldt then crossed the English Channel. Alexander's first sight of the sea at Ostend in April 1790 left the "greatest impression" on him.[14]

For Humboldt the natural scientist, his weeks in England were the most productive. They trained his comparative gaze through botanical investigations in London's Kew Gardens, where he studied flora from outside Europe, and by visiting caves in Derbyshire, further north. The journey brought Humboldt into contact with some of the luminaries of anglophone science, including Joseph Banks, the botanist and president of the English Royal Society. He was also allowed to use the library of the chemist Henry Cavendish. Humboldt's interest in chemistry steadily intensified. His mineralogical research spurred his curiosity to analyze the gases he detected both below and above ground. Cavendish extended Humboldt's knowledge with his systematic investigations of hydrogen and his measurements of gas density. They were part of a development from which modern, experimental chemistry progressively emerged in these decades. More than anyone else, the French scholar Antoine de Lavoisier epitomized chemistry as an independent scientific discipline. He explained the role of oxygen in combustion and paved the way for the modern periodic table. Humboldt missed the opportunity to get to know him personally on the last leg of his journey, which took him to Paris. Forster was keen to return to Mainz. Having arrived at the center of the revolution, Humboldt gained at least an impression of the political institutions and symbols of the new France.

Humboldt thus made the most of his student years, and his inner life in this decade proceeded at no less hectic a pace. "There is a drive in me that often makes me feel as if I'm losing my mind. And yet this unrelenting drive is essential for pushing me toward noble purposes"—so he described his situation in the fall of 1790.[15] At that time, he had not even concluded his studies. From August 1790 to April 1791, Humboldt attended the Handelsakademie (Trade Academy) in Hamburg. The journey with Forster had internationalized his worldview and placed it in a comparative perspective. His stint in Hamburg drew his attention to the importance of trade, finance, and the cash economy far beyond Germany.

From June 1791, Humboldt spent eight months at the Bergakademie (Mining Academy) in the Saxon town of Freiberg. Established only recently, the school's renown as a model of modern mining education had spread as far as America. Theoretical instruction went hand in hand with practical work underground; together, they filled Humboldt's days. Five hours' sleep were all he could afford per night.[16] Freiberg's most celebrated teacher, Abraham Gottlob Werner, took the young man under his wing. Werner was on the way to making 'geognosy,' as the study of rocks and the earth's interior was called at the time, into a science. He became one of the founders of modern geology. He taught Humboldt how to collect and classify rocks systematically and instructed him in mining techniques. Humboldt followed Werner in his adherence to Neptunism. Yet, in the years to come, he gradually moved away from this school of thought and began emphasizing the volcanic origin of most geological formations.

As the fruit of his studies in Freiberg, Humboldt published in the following year the Latin text *Florae Fribergensis specimen*, a study of Freiberg's subterranean flora, listing cryptogam plants. It combined his training in systematic classification with his passion for botany. *Florae* catalogued over 250 kinds of mushrooms and lichens Humboldt had detected in the shafts and tunnels under Freiberg. A second, more theoretical part was devoted to questions of plant physiology and chemical processes. Though hardly comprehensive and marred by errors (as we know today), this book contributed to Humboldt's fame as a founder of modern cave botany.

Physically exhausted, Humboldt ended his studies in Freiberg and thus his academic education in February 1792. From a twenty-first-century perspective, it may seem strange that he did not seek a formal qualification by writing a dissertation, for example. In the early 1790s, this rush reflected his thirst for practical experience, the offer of employment as a Prussian mining inspector, and the economic independence he would soon acquire. The natural sciences would become increasingly professionalized over the course of the nineteenth century as university curricula were streamlined. For now, Humboldt had his sights set on other challenges. Besides, the University of Frankfurt an der Oder awarded him an honorary degree in 1805—the first of seven bestowed on him by the end of his life.[17]

2

Constantly on the Move

PRACTICES AND IDEAS, 1792–1799

FOR HUMBOLDT, the period between 1792 and 1799 signified far more than a transition to a short-lived career and a phase of preparation for what became his American journey. These years were formative in both scientific and personal respects. Humboldt shared with many contemporaries the perception that time was accelerating and that the individual was exposed to dramatic change as never before: "Truly, our time resembles the strangest of times / To have gone down in history, both sacred and profane. / For whoever has lived yesterday and today in times like the present / Has already lived years: events are so crowded together." This is how the literary luminary Johann Wolfgang von Goethe, who would become Humboldt's friend, characterized the decade in 1797 in his epic poem, *Hermann and Dorothea*.[1]

Indeed, "events" and time itself were crowding together on all the continents that Humboldt came to know. In England, the incipient Industrial Revolution spawned new forms of capitalist enterprise. France deposed its king, turned into a republic, and waged war against much of Europe. The French Revolution spread to the German-speaking territories of the Holy Roman Empire. In July 1792, four months after Humboldt concluded his studies in Freiberg, Prussia joined forces with Austria against France. Tensions were rising in the French and Spanish colonial empires, too. Members of the local elites called for self-determination, undermining the insufficient reform movements of their European overlords. The Caribbean was the first region to show its revolutionary potential. In the 1790s, Haiti experienced the first successful revolution of enslaved people in modern times. Others sought to abolish the slave trade in the United States. In 1799, meanwhile, Napoleon launched a coup establishing his predominance over France.

Humboldt followed these developments with keen attention. He agreed with those who criticized absolutist rule and slavery; but he rejected terror, the guillotine, and political radicalism of all stripes. Humboldt distanced himself from ideological zealotry and attempts to abolish existing injustices by force. He never tired of advocating "moderation" (Ingo Schwarz) and pragmatic solutions. Besides, politics was not his primary concern. Revolutionary turmoil irritated him whenever it interfered with his own scientific plans. He executed these at a breakneck pace that yielded extraordinary results yet repeatedly brought him to the brink of physical exhaustion.

For four years, service in the Prussian mining administration provided the framework for Humboldt's research. In March 1792, he was appointed *Bergassessor* (inspector of mines) in Berlin; he voluntarily resigned from the post in December 1796. This occupation came with a plethora of tasks that we would today describe as state-directed resource management aimed at stimulating the domestic economy. The responsible minister in Berlin, Friedrich von Heynitz, was a pioneer of this late mercantilist policy. It sought to bring together knowledge from the technological sectors and the natural sciences while using the bureaucracy as the engine for reform. The political leadership and its officials drew on the expertise of natural scientists to improve the quality of local products and slash trade deficits.

The multitalented Humboldt was ideally suited for this reform policy. He had the cameralist know-how. His travels had allowed him to make comparisons with other regions, and he was ready to move from theory to practice. Humboldt took to his new challenges with gusto. In his first year of service, he inspected how peat, an important fuel in Prussia, was cut and how bricks were fired for use in construction. He scrutinized ceramics production to see whether it met the latest technological standards. He observed the first steam engine in Berlin's porcelain factory, experimented with plant seeds, and built an oven for chemical experiments. All this came on top of his routine inspections of widely dispersed mining sites and his extended travels in neighboring countries. Humboldt found himself always on the go. In his own words, he was "in constant movement."[2]

Humboldt's earlier study trips had familiarized him with territories west of Prussia, including in western Europe and England. His new professional position enabled him to turn his gaze to the south and east in Central Europe. In September 1792, Humboldt set out on the first of three lengthy official trips. He was tasked with studying the methods used in neighboring regions for optimizing the production of salt, a vital dietary staple. His path first led south, to Bavaria and the Austrian Archbishopric of Salzburg, then east to Vienna. Humboldt seized the opportunity to cultivate relations with Austrian natural scientists at the heart of the Habsburg monarchy. His next stop was the salt mines in Upper Silesia, which Frederick II had annexed in 1742 following the First Silesian War, as well as Breslau, today's Wrocław. He returned to Berlin via the Riesengebirge (Giant Mountains), which stretch along what is today the Czech-Polish border. Berlin did not hold him long, though.

In his first year of service, Humboldt was charged with investigating the economic situation in Franconia. In 1791, the margrave of the Franconian principalities Ansbach and Bayreuth (both belonging today to northern Bavaria) had relinquished these lands to Prussia. With that, two more territorial islands under Prussian rule arose within the Holy Roman Empire. To reach them from Berlin, Humboldt had to travel southwest through Saxon territory. He took advantage of this situation when he journeyed to Franconia in May 1793, newly promoted to *Oberbergmeister* (chief inspector of mines). Being stuck in the office was not to his liking anyway. He preferred to alternate quarters between the Franconian hills and the Fichtel Mountains, west of today's German-Czech border. Humboldt was drawn to the geological formations in these picturesque landscapes. The late medieval mines were technologically obsolete and partly lay dormant.

Humboldt plunged into his work with enormous zeal. He shuttled between the relevant mining authorities. He spent many hours down mine shafts to measure the composition of the air and crawled in tunnels to test their stability. He reactivated the Fürstenzeche mine near the village of Goldkronach and pushed for ore extraction in the district. He found the work deeply fulfilling, not least because he could connect it seamlessly with his botanical research—both above and below ground—as well as with what he called "subterranean meteorology."[3] Rediscovering primordial laws in the spatial distribution of rock strata appealed to him. Like other researchers of the time, he envisaged a 'parallelism' between geographically and chronologically distinct formations.

Humboldt also took an active interest in the welfare of workers. He looked for ways to protect their health and safety and to provide for the families of miners killed on the job. Humboldt bypassed the bureaucracy to initiate social reforms. He developed new kinds of miners' lamps, including a "light preserver,"[4] meant to help miners find their way back to daylight in emergencies. Ironically, Humboldt almost lost his own life when experimenting with this rescue device in October 1796. He also invented a "respiration machine,"[5] a mask connected to an air container that could supply miners with oxygen when harmful gases accumulated in close confines underground. Humboldt took his work on these technological innovations seriously enough to describe them at length in his book *Ueber die unterirdischen Gasarten* (*On Underground Gases*, 1799). He further arranged financial support for widows and orphans who had lost family members in mining accidents. In 1793, he drew on his own funds to establish a school of mines in Steben for boys aged twelve and above. The curriculum he designed combined basic knowledge in writing and mathematics with mining topics. Humboldt himself contributed the instructional materials. A second school in the Fichtel Mountains was subsequently added.

Humboldt enjoyed the trust of his superiors. They sent him to territories that had been redrawn in the wake of the decade's political upheavals. In early 1794, he was dispatched on a second grand tour, again to study salt and mineral extraction in neighboring regions. This time, he set out north from Berlin before curving east. First, he headed to the salt-mining town of Kolberg on the Baltic coast in Pomerania (now Kołobrzeg in Poland), then to Posen (Poznań) in the regions bordering the Kingdom of Poland. They had been assigned to Prussia in 1793 following the Second Partition of Poland and integrated into the new province of South Prussia. The busy mining officer returned via Silesia and Bohemia.

Karl August von Hardenberg, who directed the Prussian administration in the new provinces in Franconia and

FIGURE 3. Goldkronach in Franconia (2016).
© Andreas W. Daum, Private Archives.

would later become Prussian state chancellor, also de-
ployed Humboldt for his purposes. With Humboldt's
help, he attempted to introduce monetary reform in Franconia. Above all, he viewed with consternation the advance of French troops into the Rhineland. In the fall of
1794, Hardenberg instructed Humboldt to follow the su-
preme commander of the Prussian army on the Rhine and
assist him in his negotiations with the French. This was
Humboldt's first encounter with the world of diplomacy;
more were to come.

Humboldt was periodically accompanied in the Rhineland by Reinhard von Haeften, an infantry lieutenant only

a few years his junior. Humboldt had met him in Bayreuth the previous year. He loved him "dearly," Humboldt confessed in a letter.[6] Haeften was far from the first male friend to be addressed in such terms. In Berlin, Humboldt had felt drawn to the botanist Willdenow. He was similarly attracted to the theology student Wilhelm Gabriel Wegener, whom he had gotten to know during his studies in Frankfurt an der Oder, and even more so to Johann Carl Freiesleben, a classmate at the Mining Academy in Freiberg. In his private correspondence with Haeften and Freiesleben, Humboldt did not shy away from using strong language and emotionally charged images that conveyed a longing for physical intimacy, not just for spiritual or intellectual connection.

This directness undoubtedly reflected the philosophical and artistic vogue for sentiment in the second half of the eighteenth century. Men, too, should be allowed to show their feelings, indulge in their dreams, and shed their tears without inhibition. Humboldt knew Goethe's *Sorrows of Young Werther* (1774). This much-read epistolary novel showed just how dramatically masculine emotions could escalate if given free rein. *Werther* attained cult status far beyond the borders of German-speaking lands, just as circles of male friends were founded all over Europe. Humboldt intensely cultivated this form of male sensibility. The image of him as an emotionally unresponsive workaholic is a later caricature. Humboldt's devotion to his friends filled him with as much happiness as did his work.

We would do injustice to this sensitive and sensual man if we were to define him solely in philosophical or literary contexts and deny him a sexual orientation, of whatever sort. The frank homoeroticism of his letters, evident in the eighteenth century also in Frederick II, should not be dismissed as a biographical irrelevance. Brushing it aside would perpetuate the homophobic attitudes of his own day, which extended to legal discrimination. In the Prussian Civil Code of 1794, homosexuality was criminalized and could be punished severely. At the same time, we should be wary of simplistic labels. Humboldt himself makes it difficult for us to pin down his take on sexuality. He often expressed how much he felt attracted to men. Yet beyond his letters, we cannot be sure whether—and to what extent—Humboldt experienced homosexuality in the form of physical intimacy with a partner, both in the 1790s and later. After 1800, he became ever more cautious about addressing this dimension of his life. Humboldt appears to have remained open to the charms of both men and women. His love of men is no less deserving of acceptance than his wish to shelter his privacy from prying eyes. It is equally possible that his sexual orientation and its importance changed over the course of his life and that he increasingly put aside the issue of sexuality.

The sources point to other nuances. As mentioned, Humboldt never really warmed to his mother. At an early age, he had lost his father, whom he seems to have loved. The other men in the household, his private tutors, could not fill the void that opened in his life. As an adult, Humboldt frequently created obstacles for himself in his emotionally and erotically charged relationships. He pursued friendships with men who were hardly available as lovers.

Reinhard von Haeften, whom Humboldt sought "to enjoy entirely,"[7] soon married. Humboldt was torn between his desire for close male friendship and his conviction that he was destined to "wander the world . . . forever alone."[8]

In any event, Humboldt experienced considerable tension at this stage of his life, and not just in a metaphorical sense. One cause was his newfound interest in a topic that natural scientists and philosophers had discussed for a long time. Humboldt took up the question of whether there was a 'vital force,' a life-conserving quality common to all organic matter. Similar to Blumenbach's *Bildungstrieb* (formative drive), this was thought to be a force that resided in matter but could not be derived solely from material processes. Humboldt initially accepted this traditional view and made a clear-cut distinction between organic and inorganic matter. Like other scientists, he sought to design reliable tests and carried out numerous experiments on dissected animal parts and on his own body. He was interested in finding out whether, and under what conditions, organic matter remained 'irritable.' Was contact with metal necessary to stimulate reactions and trigger electrical impulses? Or did organic substances retain their irritability even after they had been separated from the organism, as the Italian scientist Luigi Galvani maintained?

Behind the concepts of galvanism and animal electricity lay an enormous diversity of explanations and philosophical subtleties. Humboldt investigated all of them tirelessly. He delved into how chemical and physiological processes interacted with electrical conductors and environmental factors to trigger reactions. Even the composition of air at the dissecting table was relevant for Humboldt. He built galvanic apparatuses, dissected frogs, rats, and birds, dripped alcohol and opium into the animals' muscle tissue, and observed how he himself reacted (with pain, unsurprisingly!) when he gave himself electric shocks. Humboldt supplemented this packed program by drawing on the work of researchers from other countries. In 1795, he conducted experiments on frog legs with the Italian Alessandro Volta, Galvani's most prominent opponent. Volta insisted that electrical impulses in the animal body could only be stimulated by contact with metals.

After three years and almost four thousand experiments,[9] Humboldt published the first volume of his *Versuche über die gereizte Muskel- und Nervenfaser* (*Experiments on Stimulated Muscle and Nerve Fiber*) in 1797. The work resembled a puzzle more than it presented a sustained argument.[10] In parts, Humboldt copied verbatim from his diary. Technical details alternated with philosophical conjectures and emotional exclamations. Humboldt juggled with various hypotheses on animal electricity, and he often admitted having gotten things wrong.

Experiments gave involuntary testimony to how cumbersome his thought process had been. Only at the end of the second volume, completed a year later, did Humboldt add the decisive piece of the puzzle. He now saw that the theory of an indestructible vital life force inherent in all organic matter, and thus his own earlier views, had been refuted. He emphasized instead that the "balance of elements in animate matter" was maintained only so long as it was "part of a whole," connected to material processes within the organism.[11]

The confusing, autobiographical character of this work shows that Humboldt's science did not correspond to an ideal of scientific research based on systematic, precise measurements, large-scale applicability, and verification through constant comparison. Since the late twentieth century, this ideal has often been termed 'Humboldtian Science,' which equates Humboldt's own research practice with an archetypal model that others later developed with reference to him.[12] Instead, Humboldt's science at this time was open-ended, situational, subjective, experimental, and almost jumpy. It lacked precision and continuity. Humboldt's science was imbued with personal feelings and insecurities in the face of considerable epistemological challenges. As a result, Humboldt's *Experiments* offered more of a collage than a closed text.

The second volume of Humboldt's complicated book ended with a quote from Friedrich Schiller's poem "Der Spaziergang" (The Walk, 1795), expressing the commitment to seek out the "stable reference point in phenomena that pass so quickly before our eyes."[13] This was no empty rhetorical gesture. It reflected Humboldt's painstaking quest to secure insights in his research while expressing his admiration for the most famous German poet of the day, besides Goethe. Humboldt's experimental work coincided with a years-long, personal exchange with Schiller and Goethe. It encompassed the topic of a vital force as well as botanical and other scientific questions. Around 1800, the natural sciences, literature, and philosophy were not yet considered separate fields of knowledge. Humboldt's brother Wilhelm helped broker the dialogue.

Wilhelm and his wife, Caroline, had moved to Jena in the Duchy of Saxe-Weimar-Eisenach in 1794. Schiller had been living there for several years, while Goethe held a ministerial position in nearby Weimar. After completing his legal training and working briefly in the Prussian civil service, Wilhelm had devoted himself to the private study of ancient Greek culture. Along with Schiller and like-minded intellectuals in Jena, he turned to classical Greece to salvage the aesthetic in an era of radical change. The terms 'neohumanism' and 'Weimar classicism' were later coined to describe this movement. Humboldt's mining districts lay only a day or two by carriage south of Jena. He could thus easily take up his brother's invitation to participate in neohumanist discussions. Humboldt visited Jena repeatedly between March 1794 and May 1797. He had already been introduced to Schiller; now the exchange with him and Goethe intensified. Both were impressed by Humboldt as a natural scientist, although for different reasons.

This perception reflected the two writers' distinct personalities. Schiller's depiction of nature, including human interactions with nature, was informed by his poetic imagination, not by empirical experience. To that extent, he remained a "disembodied author"[14] who prioritized poetic fancy and its capacity to link the presence of nature with ancient mythology. By contrast, Humboldt sought to seize hold of reality with all his senses, putting his own body on the line and deploying all the instruments at his disposal. Goethe was more open to this approach, not least due to his ongoing research interest in anatomical studies and the metamorphosis of plants. Occasionally, he even dissected animals with Wilhelm and Alexander. They later jointly visited anatomical lectures and carried out galvanic experiments. Goethe's collaboration with Alexander was spurred by his interest in finding empirical confirmation for a theory he had worked out in his imagination. When he examined the plant kingdom, Goethe proceeded from the idea of a primal form that took shape in individual specimens. Humboldt did not accept this theory, and he would distance himself from Goethe's allegiance to Neptunism in the years to come.

Still, Humboldt was eager to please both Schiller and Goethe. A yearning to be liked by his contemporaries was a constant in his life. In 1794, he thus enthusiastically accepted Schiller's invitation to contribute to his new periodical, *Die Horen* (*The Horae*). This episode says a great deal about Humboldt's ability to absorb impulses from very different scientific and artistic fields. With *Die Horen*, Schiller set out to create a forum for contemporary ideas about the classical legacy. It was a privilege for Humboldt to be the only naturalist involved in this project. His contribution appeared in the late spring of 1795 under the heading "Vital Force, or the Rhodian Genius." Schiller found the essay problematic, and it has baffled readers to this day.[15]

Humboldt explicitly presented this brief text as a literary narrative. He set it in the ancient Sicilian town of

Syracuse, a center of pre-Hellenic Greek culture. The Syra-
cusans, according to Humboldt, received a portrait from
the island of Rhodes that they struggled to understand. It
showed a godlike figure holding aloft a blazing torch and
surrounded by young men and women. The dominance
of this "Rhodian Genius" dampened their spirits and prevented them from mixing. Decades later, a second, similarly constructed picture arrived in Syracuse. Here the
genius had stood down, his torch extinguished, and the
men and women around him seemed to have shed all
restraint. They embraced each other and united in "festive
abandon." The philosopher Epicharmus succeeded in de-
ciphering the two pictures. He interpreted the genius as
an allegory of vital force. Its existence disciplined organic
matter; only its death allowed organisms to mingle freely.

Humboldt's unconventional narrative was open to interpretation in part because he liberally sprinkled the text
with scientific details and references to various philosophical traditions. "Rhodian Genius" has generally been read
as Humboldt's final effort to rescue the idea of a vital
force, an idea he had defended only the previous year in
his *Aphorismen aus der chemischen Physiologie der Pflanzen*
(*Aphorisms from the Chemical Physiology of Plants*). This
was the revised, German translation of a section of his
Florae Fribergensis from 1793.

The only literary narrative Humboldt ever wrote was,
however, part of a development in which he weaned himself off the idea of a vital force. It captured the intellectual
and social diversity that Humboldt experienced in the
mid-1790s.[16] In "Rhodian Genius," Humboldt reprised

ideas that two other authors had formulated in past issues of *Die Horen*. First, he drew on his brother's thoughts on the difference between the human sexes. According to Wilhelm, nature did not function mechanically. The female and male character worked in partnership to fulfill "the great end goal of nature, the incessant interplay of form and matter."[17] Second, Humboldt was influenced by the Swiss artist Johann Heinrich Meyer, whom he got to know in Jena. Meyer revered classical culture and was present when the Humboldt brothers met Schiller and Goethe in December 1794. Often reproduced as a drawing, subsequent generations have identified this as a seminal moment in German neoclassicism.

Meyer had been commissioned by Goethe to copy a late sixteenth-century Italian painting for exhibition in the Roman House in Weimar, the portrait of a male genius. Humboldt transposed this motif into words in his narrative. Meyer also contributed a treatise to *Die Horen* introducing readers to comparative iconography. Epicharmus used the same method. "Rhodian Genius" hence combined ideas from the different contemporary worlds— poetry and art, ancient mythology and science—that Humboldt immersed himself in. The narrative became a tribute to Jena, which had made this extraordinary synthesis possible.

Following his risky excursion into literary classicism, Humboldt was only too happy to take up the opportunity to embark on a third official trip in the summer of 1795. For the first time, it led him deep into the Alps. Hardenberg and Haeften accompanied him for parts of the

FIGURE 4. Schiller, Alexander and Wilhelm von Humboldt with Goethe in Jena (1794). Drawing by Andreas Müller.

journey, and he was later joined by his friend Freiesleben. They initially moved through the Austrian Tyrol into the territories of northern Italy. Over the next two months, Humboldt traveled extensively in Switzerland. Visits to Geneva, Bern, Basel, and Zurich were important for establishing personal contacts with Swiss naturalists. With Horace-Bénédict de Saussure at the helm, they had spent years pioneering the study of mountains and glaciers, combining geological, climatic, and botanic information

MAP 2. Locations Humboldt visited in Europe, 1769–99 and 1804–59. By Peter Palm.

in an innovative way that became a model for Humboldt. Filled with awe, he wandered through the Alpine scenery between Chamonix and St. Gotthard.

In 1796, following his Alpine expedition and another spell in the company of Haeften, to whom he still felt bound "with iron chains" despite his friend's marriage,[18] Humboldt began to distance himself inwardly from his official duties in Franconia. He dreamed of moving to Italy for a few years with the Haeftens and his brother and studying volcanoes there. Other travel destinations, from Scandinavia to Siberia, also crossed his mind. He increasingly turned his attention to the West Indies and thus the vast expanse of the Caribbean with its adjacent regions, from Mexico and Florida to the Antilles and the northern rim of the South American continent.

Besides research voyages and European colonial expansion, works of literature such as the novel *Robinson Crusoe* (1719) by the English writer Daniel Defoe had long aroused the interest of Europeans in 'exotic' places. Since his youth, Humboldt himself had been fascinated by the writings of Georg Forster and the novel *Paul et Virginie* (1788) by French author Bernardin de Saint-Pierre. *Paul et Virginie* emphasized the virtues of living at one with nature, in this case on the island of Mauritius in the southern Indian Ocean, removed from the immoral European civilization. As far as the Americas were concerned, the travel reports of scientists who crossed the Atlantic played their part in keeping Humboldt's interest alive. But he needed a trigger.

The manuscript of *Experiments* still had to be completed and various illnesses overcome. Humboldt's exhaustion was hardly surprising given his enormous workload and hectic travel schedule. His awareness that his mother lay dying may also have expressed itself in psychosomatic symptoms. To complicate matters further, political reality caught up with him once again that summer. French troops crossed the Rhine and advanced into Württemberg. Humboldt was ordered to enter negotiations with the French commander.

Humboldt's mother finally passed away on November 19, 1796. He was not by her side at the time, nor did he attend the funeral. The death of Maria Elisabeth presented him with new opportunities. Once her inheritance was settled, he found himself in possession of a small fortune of almost 100,000 talers.[19] This meant significant economic freedom for decades to come. In the late 1790s, Humboldt could initiate a journey of his own, more wide-ranging than any he had undertaken before. He no longer needed to fret about the expense or feel pressured to make plans for when he returned—or so he thought, at least. In December 1796, Humboldt resigned voluntarily from the Prussian civil service.

Early the next year, he used his newfound freedom for an extended stay in Jena, with a few side trips to Weimar. Once again, Humboldt relished discussions with his brother, Schiller, and Goethe. "Where else can one find again everything so united?" he sighed as he took his leave.[20] The alliance between neohumanism and scientific research was to be continued on the road. Together with Wilhelm's family, Goethe (who declined), and the Haeftens, Humboldt wanted to travel to Rome, a dream destination for German neoclassicists. The plan came to nothing. Wilhelm decided to go from Vienna to Paris instead. The situation in the Italian territories was growing ever more dangerous. Napoleon had led his army there in March 1796. He celebrated victory after victory over the Austrians, who had previously dominated the region. In the fall of 1797, as Napoleon redrew the political map in the south by establishing satellite states, Humboldt's travel party broke up north of the Alps.

Humboldt made the best of the situation. He immersed himself in the rich resources for transcontinental comparison that Austrian research had to offer, including the collections of the University of Vienna and the botanical holdings at Schönbrunn, the summer castle of the Habsburg royal family west of Vienna. He also conversed with the botanist and chemist Nikolaus Joseph von Jacquin, who had ventured as far as the West Indies. By his own account, he even toyed with the idea of traveling to Africa.[21]

From August to October 1797 in Vienna, then for over six months in the Salzburg region, Humboldt continued his experiments. He researched lakes and mines in the surrounding area and systematically practiced using scientific instruments. He was supported by Leopold von Buch, one of the most talented European geologists, who had studied with Humboldt in Freiberg. Together, they tested barometers, sextants with mirrors to measure angles of elevation, and eudiometers, glass tubes designed to ascertain the volume and density of gases. Such instruments had become indispensable traveling companions for Humboldt.

The waiting period dragged on. Again, Humboldt was sucked into the maelstrom of European politics, forcing him to change his plans. In November 1797, Humboldt accepted an invitation from the English Lord Bristol to join him on a tour of Egypt. He even began studying the history of Egyptian architecture. Napoleon threw a wrench in the works. By the time the Corsican made the crossing to Egypt in 1798, Humboldt had traveled to Paris to equip himself with instruments and reunite with his brother. Then the French government, known as the Directory, had Lord Bristol arrested. The Directory pursued the ambitious goal of launching a years-long circumnavigation of the world and entrusted command to the experienced Nicolas Baudin, who had previously led Austrian expeditions in the Pacific. Humboldt, whose qualifications as a roving naturalist were known in Paris, was set to join the expedition. He spent several months in "agonized suspense"[22] awaiting his departure. Again nothing. The Directory withdrew the necessary funds.

Nonetheless, his stay in Paris was not entirely in vain. Nowhere else in Europe was such a concentration of scientific expertise to be found. If anything, it had emerged strengthened from the French Revolution. In 1793, the Museum of Natural History was established, bringing the former royal collections under one roof. Two years later, the existing scientific academies were amalgamated in the Institut National des Sciences et des Arts. The bilingual Humboldt gained access to both and started networking. Soon he had forged personal contacts with the nation's leading natural scientists, including the zoologist

Georges Cuvier. He even met the legendary Louis Antoine de Bougainville, the first official French explorer to have circumnavigated the globe.

While in Paris, Humboldt also befriended the young physician and botanist Aimé Bonpland. A fellow bachelor, Bonpland wanted to travel the world carrying out scientific research. The two men joined forces and planned to cross from Marseille to Algiers before continuing to Egypt. After waiting for two months on the French Mediterranean coast, their hopes were dashed. Humboldt and Bonpland now decided to make their way to Spain instead. They intended to spend the winter of 1798–99 testing their instruments on new terrain before setting out for Africa.

Their route took them over the eastern foothills of the Pyrenees to Barcelona and Valencia, and from there to Castile, Spain's ancient heartland. In February 1799, Humboldt and Bonpland arrived in Madrid, epicenter of the Spanish global empire. They traveled in Humboldt's accustomed manner. He mapped, tested the salinity and quality of the soil, measured atmospheric pressure, and made astronomical observations, upsetting some villagers in the process. A geographical survey of Spain thus arose en route. Through his measurements, especially in the great interior plateau around Madrid, Humboldt laid the foundation for the elevation profiles of Spain he published subsequently.

Having already visited Great Britain and France, Humboldt now came to know a third European power that had for centuries operated globally and created a colonial empire in America. Yet the Spanish monarchy had suffered a

series of setbacks since the early eighteenth century, when
the French Bourbon dynasty came to the throne. In 1714,
Spain had lost its remaining provinces in the southern
Netherlands to Austria; in 1762, it was forced to cede
Florida to Great Britain. Alarmed by the crushing costs of
war, compounded by structural deficits in the domestic
economy and colonial trade, Spanish king Charles III
cautiously modernized society and the state. These Bourbon reforms extended to the colonies in America, stream-
lining their administration through the appointment of
so-called intendants and creating a kind of free trade with
the Iberian homeland.

Thanks to this reform program, Humboldt encountered
a Spain that had notched up some remarkable successes.
Spain still controlled the South American continent west
of the Portuguese possessions as well as huge swathes of
the Caribbean, including Mexico. Exports from the colonies, especially precious metals from Peru, had even
experienced a boom in the preceding decades. However,
the reform policy frequently ground to a halt as it encountered resistance both at home and among the colonial
elites. King Charles IV, who came to the throne in 1788,
faced renewed hostility from Great Britain, which sought
to block Spain's transatlantic maritime trade.

Amid these vicissitudes of European and global politics, Humboldt had a stroke of luck. In the second half of
the eighteenth century, Spain encouraged the scientific
exploration of America. Between 1759 and 1808, the crown
allowed almost sixty expeditions to pass through its colonial empire, some lasting several years.[23] Scientific interest

in plants, minerals, and other natural resources became an integral part of the strategy to make the empire's geography more amenable to political control and economic exploitation. Humboldt's timing was propitious. His training in Prussian mercantilism and cameralism turned out to be an advantage, too. Even his Huguenot descent proved no obstacle when it came to gaining favor at the Catholic court. The Spanish minister Mariano Luis de Urquijo, who belonged to the party of Enlightened reform, helped arrange an audience with Charles IV. The outcome surpassed Humboldt's expectations. The government gave him and Bonpland permission to travel through the Spanish territories in America. Humboldt received a pass and an official guarantee of protection. The Americas now stood open to him far more quickly, and with far fewer restraints, than he had thought possible when planning a journey to the West Indies.

Now events gathered pace. In mid-May 1799, Humboldt and Bonpland made their way northwest to the Spanish Atlantic coast. In La Coruna, they had to wait a few more days as British ships blocked their departure. Bearing the "most splendid recommendations and under a thousand auspicious omens,"[24] Humboldt and Bonpland boarded the Spanish corvette *Pizarro* on June 5, 1799, to set sail across the Atlantic.

3

The Interaction of All Forces

HUMBOLDT'S SCIENCE

HUMBOLDT'S DEPARTURE for America began a five-year separation from Wilhelm, his friends, and the sites of his previous research. Prussia faded into the distant background as Humboldt discovered a continent new to him. In this respect, 1799 marked a turning point in his life. But the American journey also represented a continuation of the path he had taken to his form of science over the previous decade. Humboldt's science was itself an ongoing experiment. Other researchers were still more a model for him than he was for others.

Humboldt was evolving. We should not credit him with a sovereign mastery of the most diverse fields of science and an intellectual coherence that he lacked before leaving for America and even afterward. This is a retrospective projection that doesn't do justice to his ongoing struggle to collect and generate knowledge. Humboldt was experimenting not just with mining lamps and electrical currents but with the challenges involved in synthesizing what he observed. In the words of his brother, he was "made to connect ideas, to identify linkages," and was endowed with "a rare speed in connecting the dots."[1] Humboldt's publications and his exchanges with Prussian officials and miners, natural scientists, writers, and artists had also made clear that his research was sustained by far-reaching social networks. From these relationships, and from the human desires and frustrations that surfaced in his personal friendships, Humboldt's science received its emotional charge. It addressed the heart as much as it did the mind. It remained open to error. The claim to provide precise facts was always mixed with doubt. Humboldt took the questions that arose from his mental makeup with him to America and then brought them back to Europe.

Whether on his early journeys to the Rhine, in Franconia, or soon in America, Humboldt was not just permanently on the move in geographical space. He also shuttled between different disciplinary fields, intellectual traditions, and technical applications. He turned himself into a mobile 'cloud' (to use a metaphor from our digital age) that gathered into itself an astonishing variety of information, pouring out knowledge just as quickly. Humboldt relied on the measurements provided by his instruments without becoming mechanical himself. The physical experience of his surroundings, his sensuous perceptions, and the feelings provoked in him by natural phenomena all coalesced in Humboldt's science. He also benefited from conversations on the ground and written sources. These included the existing literature in ancient and modern scholarly languages—Latin, French, Spanish, English, and German—as well as his own occasional archival research. His brilliant memory and access to his diaries enabled Humboldt to record his insights in publications composed upon his return.

This constant movement, Humboldt's intellectual openness, and the intensity with which he activated his senses allowed him to approach any given topic from different viewpoints, despite his undeniable European imprint. This combination made it possible for Humboldt to view nature in comparative and ultimately global perspectives. Through his reading, he assimilated the places he could not visit in person. Those he had visited he tucked in his memory on subsequent travels. In 1799, Humboldt left behind Rhenish basalts, Franconian mine shafts, Austrian salt mines, Swiss Alpine lakes, and Spanish mountain ranges. But they all floated up from memory as he gazed on American landscapes.

From 1796, Humboldt kept himself informed about the many expeditions that had already explored Spanish America. His science emerged from both established traditions and fresh impulses. Humboldt's journey belonged to an era of research expeditions that extended from the mid-eighteenth century deep into the nineteenth century. Humboldt was familiar with the Amazonian journey of the French explorer Charles-Marie de la Condamine and the expedition mounted by his compatriot, Pierre Bouguer, in today's Peru. He admired the works of Spaniard José Mutis on South American flora and fauna. He was eager to learn more about the travels that had brought the

Italian Alessandro Malaspina, sailing under the Spanish flag, as far as the Pacific region. While in Vienna, he had personally quizzed Jacquin, the Austrian traveler to the West Indies. In Madrid, he used the botanical garden and the natural history museum to inform himself about the findings of Thaddäus Haenke, a Bohemian naturalist whose travels to South America succeeding generations have unjustly bypassed.

Humboldt would later be celebrated as a second Columbus and new discoverer of America. Such simplistic tropes not only reflect a colonial worldview, they were also far from his mind when he set sail across the Atlantic in 1799. Humboldt was not venturing into unknown territory. Nowhere was he the 'first.' Still, his journey was more individualistic, and freer in its choice of route, than those of his predecessors. He also came better prepared than most, endowed with a more refined and nuanced gift for observation. Humboldt brought an unparalleled intensity to his mission to study interactions within nature, and between the natural and the human world. Today, we would call this approach ecological at its core. Was this still the ancient aspiration—given new life by his teacher Lichtenberg, by Georg Forster and Goethe—to grasp nature as a whole? Or did it anticipate Humboldt's attempt to understand nature with modern methods and empirical proofs as a new 'cosmos,' the title he gave his magnum opus half a century later? We must take care not to leap to premature conclusions about his earlier works based on his later ones.

Humboldt's ideas were not yet fully formed, and he was not the sole founder of modern ecology. Other contemporaries such as the French geologist Jean-Louis Giraud Soulavie, the German geographer Eberhard August Wilhelm von Zimmermann, and Humboldt's friend Willdenow reflected in their studies on the complex inter-relationship between climate, geography, plants, and animals. The impact of climate and nature on human society in different geographical settings was a prominent theme among Europe's philosophers, too.

It was thus more with a goal than with any concrete agenda in mind that Humboldt asserted upon departing from Europe in June 1799: "The interaction of all forces, the influence of inanimate creation on the animated worlds of flora and fauna—that is the harmony on which my eyes shall constantly be fixed."[2] This was an audacious vision for one man to seek to realize. Humboldt wanted to integrate intellectual traditions and his own field research in an empirically saturated, comprehensive view of nature. Wilhelm, more acutely aware of his brother's idio-syncrasies than anyone else, sensed the danger (as did Humboldt himself) that Alexander would overreach himself and get bogged down in details.

Adding to this concern was the often erratic way in which Humboldt varied the imperative of late eighteenth-century natural philosophy to unite reason with nature. Along with all naturalists of his time, his chief interlocutor was Immanuel Kant. Humboldt was profoundly influenced by Kant's attempt to provide a systematic basis for physical geography as the totality of natural phenomena and to establish geography as an independent spatial science. Yet adapting Kant's critical philosophy to his own ends proved difficult. And unlike his brother Wilhelm, he could not afford to spend months poring over books.

Humboldt wrestled with the epistemological challenges of his self-imposed goal to strive for a holistic conception of nature and with the attempt to arrive at objective results through experiments and measurements. He struggled to avoid conflating reality with the sum total of empirical data provided through direct observation. From Kant, Humboldt had learned that it was equally important for the researcher to be aware of the a priori conditions of all experience, that is, guiding ideas, mental constructs, and hypotheses. But in the 1790s, he often foundered in his attempts to strike a balance between empirical analysis and his ideas. As a consequence, he was repeatedly forced to revise research results that he had initially seen as objectively valid.

Humboldt therefore took with him to America the challenge of arriving at a coherent way of thinking that would allow him to marshal scattered information and creatively synthesize it within a comprehensive framework. How could he reconcile the need for objectivity with his subjective perception? How could he grasp the "interaction of all forces" without lapsing into encyclopedic compartmentalization? How could he collate the data he would gather into general statements? One thing was certain: Humboldt would have his hands full in America.

4

Gaining a Picture
of the Whole

THE AMERICAN JOURNEY,
1799–1804

THE AMERICAN JOURNEY that Humboldt embarked on in June 1799 proved epic in both scale and duration. It lasted over five years, far longer than expected. Humboldt was repeatedly forced to change his plans. By crossing the Atlantic, sailing the Caribbean (which he called an "American Mediterranean")[1] and the Gulf of Mexico, and charting a sea route along the eastern rim of the Pacific, he became acquainted with different oceanic zones. His excursions inland began on the Canary Islands, close to the northwest coast of Africa. These had long been a Spanish imperial possession. He then traveled to areas in South and Central America.

The regions he traversed were administered at the time by the Spanish crown as the Captaincies General of

Venezuela and Cuba as well as the Viceroyalties of New Granada, Peru, and New Spain. Independent states would emerge in these territories only in the decades following Humboldt's return. For ease of reference, I will apply today's political borders and use the modern names of countries and states. Initially, then, Humboldt traveled through Venezuela. The soon-to-be legendary excursion on the Orinoco River into the continent's interior led partly along the current border to Colombia. He then crossed over to Cuba, from where he sailed to Colombia before traveling inland to Ecuador and Peru, mostly following the mountain range of the Andes. From Peru, Humboldt reached Mexico via the Pacific. He spent almost a year there before making a second trip to Cuba. Humboldt next left the Spanish colonial empire to visit the eastern seaboard of North America. In July 1804, he crossed the Atlantic and returned to Europe.

Humboldt was aware that he came to America as a European whose outlook had been shaped by his Prussian upbringing and training. Notwithstanding the support of the Spanish crown, he traveled on his own initiative in a private capacity. Humboldt and Bonpland aimed to achieve together what other research expeditions had spread over much larger teams with politically motivated interests—navigating difficult terrain, constantly taking measurements, improving maps, drawing on paper what they saw, recording their experiences, and all the while expanding their knowledge of botany, geography, climate, zoology, and astronomy. Even this list is incomplete. In addition, Humboldt slipped into roles that we would

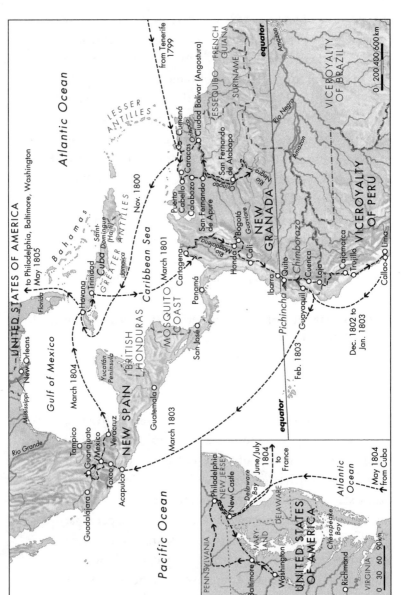

MAP 2. American journey, 1799–1804. By Peter Palm.

today describe as those of an economist, sociologist, demographer, anthropologist, ethnologist, and cultural historian. He endeavored to understand human societies through their cultural artifacts. Humboldt wanted to decipher the mutual influence between human beings and their natural habitats. He wished to know how societies are shaped by the natural resources they use and often exploit.

The beginning of the journey already showed Humboldt putting his guiding principles into action. In mid-June 1799, he spent a full five days on the Canary Island of Tenerife, west of Morocco, a much frequented stopover for Atlantic travelers. Its volcanic structure and flora had long attracted the attention of natural scientists. "Saw, felt, and asked about so much," Humboldt summed up in his diary.[2] Together with Bonpland, he visited the botanical garden and admired a colossal dragon tree in Orotava. From there they climbed the Pico del Teide volcano, which towers at over 3,700 meters above sea level. Blasted by icy winds, Humboldt ventured to the crater's edge. His observations confirmed the volcanic origins of basalt. Coming almost a decade after his visit to the Rhenish basalt formations, this was an important step away from his earlier, Neptunist views. Humboldt immediately noted the differences in vegetation zones along the mountain ridge and the social inequalities on the island. The researcher felt committed to the ambitious goal that framed his approach to what he encountered. From the many details he documented along the way, a "picture of the whole" should emerge.[3]

It was on the next stage of his transatlantic voyage that Humboldt first set eyes on the Southern Cross, as sixteenth-century European travelers had labeled this constellation. His priority during the crossing was to describe as precisely as possible phenomena he had never encountered before, for example, the oxygen content of the sea air and the characteristics of so-called flying fish. On July 16, 1799, the *Pizarro* anchored off Cumaná in Venezuela, east of Caracas. For the first time, Humboldt and Bonpland set foot on the South American continent.

Immediately, they began exploring their new environment, which included the shale deposits and gold mines of the surrounding mountains and the saltworks of the Araya Peninsula. Here, as throughout the journey, Humboldt's training in cameralist practices paid off. He quickly recognized whether natural resources could be practicably and profitably exploited; that he had learned in Prussia and Europe. But the visit to the Cueva del Guácharo, the biggest stalactite cave in the continent, astonished even an experienced speleologist like Humboldt. The local guides left him and Bonpland alone with the shrieking oilbirds that nested there. Once a year, the indigenous killed thousands of these mythically transfigured birds to harvest their fat. Bonpland bagged two of the oilbirds, while Humboldt made drawings of their anatomy.[4] Dramatic moments would punctuate their travels from then on. Even before they had left Cumaná, the Franco-Prussian duo experienced a solar eclipse, an earthquake—and an attack on the beach.

In Cumaná and Caracas, Humboldt could observe the transformative processes that had the Spanish colonial empire in their grip at the time. Transport routes from the two Americas, the Caribbean, the Atlantic, and Africa all converged on the north coast of Venezuela. European powers had been infiltrating these regions since the late fifteenth century. Soon, widespread intercontinental traffic in humans and animals, raw materials, precious metals, plants, technologies, and diseases arose between them. Humboldt understood these transfers and the mutual influences that ensued, long before historians coined the term 'Columbian Exchange' to describe them.

He realized, too, that this transcontinental exchange was made possible by the cruel deportation of millions of people from Africa and their enslavement and thus came at huge human and ecological costs. While the postcolonial critique of Humboldt, articulated with verve since the late twentieth century, is a necessary corrective to an idealized, glorifying image of the Prussian explorer, it misses at times the complexity of his thinking and observations. Humboldt did not see South America exclusively with "imperial eyes" that made him construct the "image of a primal nature." His later descriptions drew on many inspirations and motivations, not only using the "Euro-expansionist process" as a reference point or deterritorializing indigenous people.[5]

Already within the first six months of his voyage, Humboldt astutely recognized how the 'Columbian Exchange' had caused massive structural inequalities. It served Europe's growing demand for consumer goods derived from the cultivation and processing of tobacco, sugarcane, coffee, and cotton. The selfish interests of European colonial overlords and entrepreneurs drove the unsustainable pillaging of resources on the ground. Humboldt understood the negative consequences that resulted from the clearing of land for agriculture and deforestation. Above all, the mainstay of the colonial economy, the slavery system, was coming under considerable strain. Between 1701 and 1810, the European colonial powers deported almost six million people from Africa to exploit them as a labor force in the Americas.[6] But the measures required to keep them under control, and the concentration of the plantation economy on individual products, proved expensive and ultimately inefficient. Late mercantilist economic policy in Prussia had trained Humboldt to diagnose such problems.

Humboldt's humanitarian convictions influenced how he perceived the world of enforced labor. In the coastal regions of Venezuela, he witnessed how men who had been brutally captured and kidnapped were put up for sale and flogged. The fact that the Spanish colonies enslaved fewer people than those under Portuguese control and the British and French Antilles did nothing to diminish Humboldt's outrage. In his travel diaries, and only slightly toned down in his later writings on America, Humboldt unequivocally condemned slavery as the "single greatest evil to torment humankind."[7] Still, contradictions remain. In the Americas, Humboldt got to know both owners and traders of slaves, and some enslaved people served him on his journey. He came from the Europe of the Enlightenment and the French Revolution of 1789. Now, thousands of miles away, Humboldt realized that their ideas were both embraced and repudiated by many in the Caribbean and in South America. The epicenter of political upheaval initially lay in Hispaniola, the second-largest island in the Antilles after Cuba. Its western third, later and still today known as Haiti, belonged to the French colonial empire as Saint-Domingue. The revolutionary emancipation of enslaved people in Haiti from 1791 electrified the western hemisphere.

Humboldt was anything but alone in taking a stand against slavery; protests grew ever louder in Europe and the colonies. Although he never became an activist in the abolitionist movement, his was still a bold position. Humboldt turned on its head the colonial logic that glorified the oppressors for having civilized the oppressed. The Spaniards, he insisted, had learned far more from the natives of America than they had taught them. Nonetheless, Humboldt did not see himself as a revolutionary. He sympathized with the liberals who linked freedom and property rights and placed their hopes in the reforms initiated by the Bourbon kings. Ultimately, he sensed that the path of historical development was leading to national independence for the exploited territories.

Humboldt registered with great sensitivity just how complex and fractured Spanish colonial society was. Besides enslaved people and those freed, he also met descendants of America's pre-Columbian populations. He studied the customs and mores of these so-called Indians down to their linguistic peculiarities, often recalling the interests of his brother in Europe. He voiced his outrage that the Spanish invaders had killed so many of their ancestors, while their descendants continued to suffer discrimination and exploitation.[8] He also differentiated between the sections of the populace that enjoyed access to education and social mobility and the less privileged, mixed-race 'mestizos' who were denied it. Humboldt clearly saw that the so-called *mulattos*, those of mixed 'white' and African ancestry, and the *zambos*, who descended from both indigenous and Africans, were especially disadvantaged.

Furthermore, he witnessed firsthand the differences within the Spanish administration and the ethnically Spanish, South America–born elite. Some of the representatives of this Creole class struck him as pompous and corrupt, others as well-informed and receptive to new ideas. The longer Humboldt's journey lasted, the more he encountered the many scattered groups of people who made this ethnic and social kaleidoscope even more diverse. He met miners from German-speaking countries, French soldiers, and descendants of the mostly exterminated Caribbean indigenous population. The various representatives of the Catholic Church were ubiquitous, and he frequently came into contact with missionaries. Following the expulsion of the Jesuits from the Spanish empire at the behest of Charles III in 1767, monks from the divisions of the Franciscan order dominated. In his diaries, Humboldt was vocal in his criticism. Most missionaries were immoral, despotic in their dealings with enslaved people and 'Indians,' fraudulent in their trading practices, and addicted to luxury. Still, he needed them as much as local, mostly indigenous helpers.

Humboldt's science turned out to be both an individual and a collective enterprise. This became apparent in February 1800, when Humboldt and Bonpland set out for the interior of what is now Venezuela. Box after box of measuring equipment, books, and containers filled with plants that they collected along the way had to be transported. The helpers served as porters and found shelter for the group when they came across impassable rivers. Their indigenous knowledge proved invaluable as they drew on communal traditions and an intimate familiarity with the land. The latter was necessary to shoo jaguars away from camp, find their bearings in the seemingly trackless wilderness, and locate food and water. The menu ranged from roast monkey to ants. Indigenous and *mulattos* also helped in the attempt to find a modicum of protection from the ongoing mosquito plague.

Mosquitoes in Central and South America had long become "key actors" not just in nature but also in the Europeans' geopolitics, which had a massive impact on natural habitats.[9] The colonial system destroyed ecological balances, increased the number and diversity of insects, and promoted the transmission of diseases. In the tropics, Humboldt and Bonpland often buried themselves up to their heads in the ground at night or crawled into low, smoky ovens to avoid being covered in bites. Genuine relief was impossible. Fever, nervous irritation, skin rashes, and itching were an everyday reality. It was also annoying for Humboldt when dense swarms of mosquitoes obscured his vision of the starry sky and the plant world.

From Caracas, Humboldt, Bonpland, and their helpers moved west to survey the mountain lake of Valencia and the harbor town of Puerto Cabello. Then came the decisive change in direction south through the plains of the Llanos to San Fernando de Apure, a base for Capuchin missionaries. Near Calabozo, Humboldt's galvanic interests caught up with him. A local man surprised the author of *Experiments on Stimulated Muscle and Nerve Fiber* with his self-built electric machine. This anecdote revealed in a nutshell that the Hispanic world had generated its own interests in the questions Humboldt was pursuing, while information and technology were spreading over the globe. He had spent years in Europe investigating whether and how animals could have an electric charge. In Calabozo, he observed how electric eels attacked horses midstream and gave them electric shocks. Humboldt tried the experiment on himself and felt considerable pain. That, too, almost had a symbolic meaning. The European researcher was not a neutral observer in South America but responsible for his experiments' consequences.

From San Fernando de Apure, Humboldt and Bonpland continued their journey by boat. Over the next three months, they covered more than 2,000 kilometers on waterways. In early April 1800, they reached the Orinoco River via the Rio Apure. The group followed it upstream around 400 kilometers to the south, penetrating deep into the wettest zone of tropical South America. This route was of political interest as it allowed the border zone with Portuguese Brazil to be mapped more precisely. Humboldt was appreciative of the pioneering work the Spanish border expedition of 1754–61 had carried out, but he did not enter territory controlled by Brazil.

The main objective of the journey up the Orinoco stemmed from Humboldt's desire to think in terms of continent-wide connections and to demonstrate these empirically. He wanted to prove that South America's waterways formed a single, interconnected system from the Andes to the Amazon. For Humboldt, the Orinoco was the "key" to the continent, as he noted in his diary.[10] This focus required him to advance as far as possible into the Orinoco's source region while extending the route from there—by adding a short overland leg—to the upper course of the Rio Negro, which flows into the Amazon, as well as the Rio Casiquiare, which connects the Orinoco and the Rio Negro. From a bird's-eye view, this turn resulted in a kind of loop in the southern part of the tour.

Humboldt succeeded in proving his theory, albeit not without considerable strain. The Orinoco voyage had barely begun when their boat almost capsized. Books, drawings, and dried plant specimens all went overboard. Then the local helmsman went on strike. Humboldt and Bonpland had to switch boats and navigate the dangerous rapids at Atures and Maipures. They faced rain and dreary swamplands, constantly plagued by mosquitoes and other insects that crawled into their mouth and nose and even settled in their finger joints. To make matters worse, Humboldt's cherished travel barometer broke.

At the end of May 1800, Humboldt and Bonpland turned back. This time they stayed on the Orinoco until they reached today's Ciudad Bolívar. Humboldt decided to return to the north coast of Venezuela. Then there was the question of what would become of the plant collections they had amassed and the live animals they had captured along the way, including several monkeys. Much was shipped across the Atlantic and went missing at sea. It is almost a miracle that Humboldt's ink-penned diaries survived the voyage. They became a treasure trove for their author and his colleagues upon his return—and remain so to this day, now successively accessible in new editions and in digitalized format.[11] Humboldt did not keep a diary in the classic sense, one that chronicled the journey in entries written at regular intervals. Instead, he combined travel depictions with tables of data, excerpts from his reading, more systematic observations on specific topics, drawings, and other scattered texts. From 1802, he increasingly shifted from German to French in his writing.

On November 24, 1800, Humboldt and Bonpland left Venezuelan territory to sail to Cuba. They had no plans to revisit South America; Humboldt was occupied with other ideas. He contemplated sailing back to Europe via Cuba and Mexico, then across the Pacific, the Spanish Philippines, and India. Yet events took a different course. Humboldt spent almost three months in Cuba. The intellectual culture in Havana appealed to him. His visits to the surrounding *haciendas*, the plantation estates, left him with the impression that the enslaved people were treated better on the sugar-producing island than elsewhere in the Caribbean. Humboldt collected extensive demographic data. His recommendations for improving boiling methods and ovens in sugar production gained a hearing in the following years; he had studied similar issues in the 1790s in Europe.

Meanwhile, newspapers from Europe were reporting that the French explorer Baudin was on his way to the west coast of South America as part of his latest Pacific voyage. Humboldt felt bound to his earlier promise to Baudin and changed his plans. He and Bonpland deposited part of their plant collections in Cuba, and in early March 1801 they set off again for the South American continent. After they were waylaid by a sea storm off the coast, they finally reached Cartagena in present-day Colombia. During that journey and on other voyages by boat, Humboldt never became seasick. The two travelers reached the southern town of Honda via the foothills of the Rio Magdalena, gaining access to the Central Andes and mines in the Cordilleras. Humboldt's experiences in Freiberg and the Franconian mines accompanied him wherever he went.

The next stop on their itinerary was Bogotá, capital of the Viceroyalty of New Granada, perched over 2,600 meters above sea level. Here Humboldt finally had the opportunity to meet his illustrious predecessor, the nearly seventy-year-old Spanish botanist and mathematician José Mutis. Humboldt followed his tried-and-tested procedure from his time in Central Europe, combining collegial exchange and the study of archival sources and scientific literature with extensive field research of his own, for example, in local salt and silver deposits. A baggage train of twelve mules escorted them further south over the Andes. This was a perilous undertaking; later, some of their belongings would go missing on such treks. As 1801 made way for 1802, Humboldt arrived in Ibarra, which in 1830 would become part of a newly independent Ecuador. Francisco José de Caldas now joined the party. Born in South America, Caldas was an ambitious autodidact who had constructed some of his own instruments. Humboldt particularly appreciated his talent for astronomy.

As on previous stages of their journey and in Europe, field research was a team effort. Humboldt always sought out skilled naturalists for his travel companions. He needed research assistants, laborers, and porters. Humboldt relied on the knowledge amassed since the sixteenth century by colonial researchers, earlier missionaries, and Spanish conquistadores as well as by French, Spanish, and Creole naturalists of his own time. The spectrum ranged from Pedro de Medina, Andrés Urdaneta, and Pedro Cieza de León to Charles Marie de La Condamine, Mutis, and Caldas. Whether he and later historians of this knowledge transfer did enough to acknowledge the non-European sources of today's ecological and environmental understanding has meanwhile become the subject of lively debate.[12]

Along with Caldas, Humboldt and Bonpland arrived in early January 1802 at the city of Quito, located at an even higher altitude than Bogotá. By now Humboldt had become something of a celebrity. Awaited with a mixture of curiosity and admiration, the man from Berlin was gawked at, interviewed, and celebrated upon his arrival. His adventures as a private scholar traveling with so large a retinue lent him a whiff of the exotic. In Quito he was deluged with visitors and invitations. Far more important to Humboldt, however, was that Quito offered him the unique opportunity to investigate vulcanism in one of the most spectacular mountainous regions on earth.

Humboldt had long been drawn to the idea of systematically exploring the horizontal and vertical dimensions of the natural world and investigating their interrelationship. No force in the universe, he had written in 1799, could "be conceived independently of any other. . . . In nature there is neither above nor below."[13] Precisely to demonstrate this continuum, Humboldt felt compelled to scale mountain peaks and descend to the bowels of the earth. His exploration of caves and mines in Europe and his forays in the Swiss Alps had prepared him well. He had already used his ascent of Pico del Teide on Tenerife and Puracé in Colombia to describe the gradation of geological, botanical, and climatic zones on volcanic mountains. The Andes offered him new possibilities.

Climbing to the crater rim of a volcano was Humboldt's goal. From there, he could peer into the geological depths, test sulfur fumes, measure fluctuations in temperature inside and outside the crater, and reexamine the old question of the origin of rocks. The results in Ecuador confirmed for Humboldt that Plutonism offered the most plausible hypothesis. He realized that he was on one of the volcanic belts of the Andes linked by underground magma flows. The fumes on the surfaces of the volcanic mountains showed him that lava was formed by the melting and intermingling of metals and alkaline earths. He was unaware of the primordial drifting of continental plates; that only became common knowledge much later. Yet Humboldt anticipated the insight that not all earthquakes result from these deep geological faults.

Humboldt's ascent of Mount Chimborazo south of Quito on June 23, 1802, looms large in the memory of posterity. While Humboldt was unable to reach the crater rim, he set an altitude record of around 5,500 meters for European travelers.[14] He later took considerable pride in this feat, even though locals had climbed higher peaks in the Andes before him. Chimborazo was the last of five volcanic mountains that Humboldt scaled in the space of three months, with heights ranging from around 4,700 to 6,200 meters. The first was Antisana southeast of Quito, followed by Pichincha to the west and Cotopaxi and Tungurahua further south. On Tungurahua, too, fissures in the rock prevented him from reaching the summit. Humboldt's most important achievement was his persistence in understanding these climbs as a series of interlinked experiments. Repeating and comparing such experiments had become a core part of Humboldt's science.

In his galvanic experiments during the 1790s, Humboldt had put his body on the line in his pursuit of scientific research. On the Andean volcanoes, he paid the price for this connection to his own sensations by pushing himself to the brink of physical endurance—and sometimes beyond. Modern high-altitude mountaineering gear such as goggles, helmets, and oxygen masks did not exist.

FIGURE 5. Alexander von Humboldt and Aimé Bonpland in the plain of Tapia at the foot of Mount Chimborazo. Painting by Friedrich Georg Weitsch (1806).

Humboldt almost lost his life on more than one occasion. He suffered from chest pains and respiratory difficulties as the atmosphere grew ever more rarefied. His gums and lips started bleeding, and his eyes became bloodshot as his vessels burst. His hands tore open, his feet were rubbed raw. Then there were the "crushing headaches,"[15] heart complaints, and blinding glare from the snow, which in Humboldt's view reflected the sun more intensely than in the Alps. On Pichincha, Humboldt fainted after an attack of vertigo and had to be rescued by his helpers.

Humboldt noted how, under such extreme conditions, sensory experience opened new spaces for the imagination to perceive nature's beauty and appreciate it as a "magnificent spectacle."[16] This spectacle deserved as much to be measured with instruments as to be recorded in images and texts. Humboldt appeared as an actor on the stage of this spectacle, while later adopting the role of a spectator whose impression was important for evaluating the performance. Measurement, physical exertion, and aesthetic pleasure went hand in hand.

Even more than the voyage on the Orinoco, Humboldt's experiences in the Andes made clear that he was moving toward a holistic view of nature that placed value on aesthetic perception. He was touching on ideas that, back in Europe, were coalescing into the Romantic philosophy of nature. To be sure, Humboldt yielded often enough to his obsession with measuring things. The German writer Daniel Kehlmann exaggerated this trait in his bestselling novel, *Measuring the World* (2005).[17] But Humboldt knew that the human desire to arrive at "general ideas"[18] went far beyond what his instruments documented. Schiller, who died in 1805, would have been pleased to read this in Humboldt's diaries.

The ascent of the volcanoes was followed by a gradual, winding descent into southern climes and today's Peru. After Quito, and until their return to Europe, Humboldt and Bonpland were accompanied by Carlos de Montúfar. The Creole aristocrat would later distinguish himself in the Ecuadorean struggle for independence. In Cuenca, the locals put on bullfights in Humboldt's honor. In Loja, he could study the so-called China tree, whose bark was used to combat fever. Humboldt first caught sight of the

Pacific south of Cajamarca. He then made his way along the coastal route to Lima. In southern Ecuador and Peru, Humboldt pored over the surviving traces and traditions of the Incas. Upon his return, he provided his brother with a detailed report on their many languages; this was Wilhelm's area of expertise, after all.

Through his research, Humboldt challenged the Spanish colonial empire's claim to have bestowed the blessings of civilization on the continent's first peoples. He showed an appreciation of these earlier cultures even as he imposed his European scientific practices on their very different traditions. Ignoring the protests of his native companions, for example, Humboldt investigated skulls at an old burial site, the cave of Ataruipe. He removed some of his finds and had them shipped off to Europe.[19]

In October 1802, Humboldt and Bonpland reached Lima, the southernmost point of their American itinerary. Over the course of his entire journey, Humboldt moved between the 40th parallel north and the 12th parallel south, roughly. By carrying out regular geomagnetic measurements, he provided more precise evidence that the polar magnetic fields weakened toward the equator. In the port city of Callao, the travelers finally boarded a Spanish frigate to take them north through the Pacific. Humboldt's onboard measurements confirmed the existence of a large, cold-water ocean current in the eastern Pacific. It was later named the Humboldt Current, much to his embarrassment; he knew that he was far from the first to have drawn attention to it. In Guayaquil, Humboldt and his companions spent a few more weeks again exploring the

inland regions of Ecuador. They then sailed east past the
Galapagos Islands to the Viceroyalty of New Spain, today's
Mexico. On March 22, 1803, the group arrived in Acapulco.
Over the next three weeks, a pack of mules assisted them
as they made their way via Taxco to Mexico City.

The Mexican leg of his journey differed from its earlier
stages. Humboldt spent an unusually long time in the region, almost a full year. With Mexico City, he gained a
central base where he could settle for months. He valued
its urban culture and the opportunities for scientific study
it afforded him. At the School of Mines, Humboldt met
two colleagues who had also studied in Freiberg. Academic pathways and the dispersal of graduates played
their part in the internationalization of knowledge. Hardly
anything escaped Humboldt's attention. He registered
birth and death rates, Mexican beard styles, pre-
Columbian writings, and the architectural peculiarities of
the Pyramid of Cholula, which he compared with Egyptian monuments.

From Mexico City, Humboldt struck out into the surrounding regions. He was keen to investigate the diverse
geological formations of the Sierra Madre and to inspect
the extraction of natural resources, examine drainage projects, and study commercial structures. The recent convert
to Plutonism could not resist an excursion to the twin
volcanoes of Popcatépetl and Izaccíhuatl, southeast of
Mexico City. His stay in the northwest region of Guana-
juato, the third-biggest city in the Spanish colonial empire,
was especially taxing. Humboldt again sought to arrive at
a cross-section of nature by measuring atmospheric

pressure at high altitudes and venturing far below the surface of the earth. His landscape descriptions increasingly became a composite of texts, images, and topographic maps. The profiles that resulted combined horizontal and vertical dimensions. Humboldt even experimented with a formulaic language (pasigraphy) that aimed to condense detailed information into memorable symbols.

Humboldt and Bonpland left Mexico in early March 1804. From Veracruz, they crossed the Gulf of Mexico a second time to arrive in Havana. This time they stayed only six weeks in Cuba. Humboldt used them to gather the remaining data for his planned geography of the island. By now he was fluent in Spanish. A detour to the United States had not been planned but came about thanks to the American consul in Havana. At this final stage in Humboldt's journey, disaster was narrowly averted when his ship foundered off the Bahamas.

Humboldt and his companions finally sailed into Delaware Bay on May 20. They made for Philadelphia, the largest city in the United States and its scientific capital. Humboldt enjoyed exchanging ideas with members of the American Philosophical Society, the oldest learned society in the United States, founded by Benjamin Franklin. Subsequently, he met the leading representatives of the U.S. government in Washington. President Thomas Jefferson repeatedly made time for him. Humboldt seems to have steered clear of the topic of slavery. Jefferson, who knew Europe firsthand and embraced enlightenment ideas, was a slaveholder on his estate in nearby Virginia.

Humboldt's gallant appearance in the United States and his avoidance of political controversy were in the interests of both sides. With the Louisiana Purchase the previous year, the American government had opened a gateway to further expansion to the west. In 1803, France sold the United States an area of more than 1.4 million square kilometers, stretching from the Gulf of Mexico to the Rocky Mountains and north past the present-day Canadian border. It encompassed vast swathes of what is now the American Midwest, bringing the young nation into direct contact with the Spanish colonial empire in the west.

Although Humboldt had never set foot in these border regions, Jefferson and his colleagues were eager to tap him for information about New Spain's demographics, natural resources, and economic prospects. These were relevant for the next phase in the United States' westward expansion. Humboldt obliged; he too could profit from the exchange. Over the following decades, he corresponded with a growing number of scientists, writers, and travelers from the United States. They provided him with a North Atlantic supplement to the information on Central and South America he had been gathering for years.

Turning his back on the "magnificent world" of the Americas was not easy, as he wrote his old friend Freiesleben. The journey had made him "healthier, stronger, more hardworking, even more cheerful than ever."[20] Still, Humboldt longed to return to Europe. On June 30, 1804, the illustrious travel group embarked on the Delaware to make the passage across the Atlantic to France.

5

Delivering to the Public

IN PARIS AND CIVIL SOCIETY, 1804–1827

HUMBOLDT AND HIS COMPANIONS arrived at the French Atlantic coast early in August 1804. After five years abroad, they again set foot on European soil, setting out from Bordeaux for Paris. There Humboldt met his sister-in-law, Caroline. Her impression of him spoke volumes. Alexander, now almost thirty-five, had barely changed in his manners, though he had gained weight! Relishing his fame, he was eager to go everywhere and see everyone in Europe. Overwhelmed by his breathless reports and boundless energy, Caroline felt her "head spinning like a top."[1] Consciously or not, she was summing up the mood of the age. Central Europe was gripped by forces that many found both exhilarating and bewildering. Old orders had been swept away and new political structures established in their place.

In the years of Humboldt's absence, Napoleon had extended his hegemony over continental Europe. Wars and armistices alternated with stupefying regularity. The Holy Roman Empire was on its last legs. France had occupied the areas west of the Rhine. The empire reacted in 1803 with a far-reaching territorial reorganization. Ecclesiastical principalities, along with other territories, were secularized and assigned to the larger German states. Prussia was among the beneficiaries. A few months after Humboldt's return, Napoleon crowned himself Emperor of the French. In 1806, he took the war to Prussia and the Russian Empire. Humboldt personally experienced French occupations and the emergence of a German national consciousness. He saw how regime changes went hand in hand with the drawing of new borders and the political mobilization of the civilian population.

The same period witnessed an intensification of social and intellectual changes underway since the eighteenth century. The traditional, hierarchical order dominated by the landed nobility was losing ground. Social groups that we would today call middle-class were on the rise. This *Bürgertum* had an economic wing, comprised of businessmen and merchants, and an educated wing, drawn from professionals such as doctors and lawyers, intellectuals, and other university graduates; the latter included the bureaucracy and Protestant clergy.

This heterogeneous middle class was united in its aspiration for social mobility, respect, and a greater say in politics. It placed a premium on work and individual achievement—and on *Bildung*, the most important key to social success besides property. The middle-class citizens, mostly urban *Bürger* dominated by men, demanded access

FIGURE 6. Alexander von Humboldt (1807).
Drawing by Frédéric Christophe d'Houdetot.

to educational opportunities that would allow for individual self-realization. Women were clearly at a disadvan-
tage here. But the idea of *Bildung*—a general, humanistic
education that stood to benefit the community as a
whole—was becoming increasingly important. To make
it a reality, an expanded public sphere that transcended
class divisions would have to be created and nurtured. If

the effort to subject everything to critical discussion was to move beyond the elite Enlightenment salons, which Humboldt had frequented in Berlin, communication skills and professional competence would be crucial.

Humboldt thus found himself in a transformed political and social climate, and he faced new challenges. Would his research suffer amid the ongoing turmoil in Europe, especially given the concentration he needed to work through the findings of his years-long journey? Could he still pursue his quest for a knowledge that transcended boundaries in an age that was violently imposing new ones? Would he be able to reconcile his love of French culture with the looming military showdown between Prussia and France? How would he, an unconventional scion of the Prussian nobility, arrange himself with middle-class interests and values? Humboldt had previously led a footloose, relatively independent life. Having always avoided political commitments, would he now be forced to choose sides? Humboldt's answers to these questions show him adapting creatively to changed circumstances. He became a kind of roving ambassador for the sciences, often juggling different commitments and navigating between all cultural and political fronts.

In the first book he published following his return from America, the French-language *Essay on the Geography of Plants* (1805), Humboldt made his target audience clear. He was writing not for any state but for *le public*, "the public."[2] This French term summed up what belonged together in German-speaking countries, too: critical readership and, more broadly, the public sphere—*Öffentlichkeit*

in German—as the most important sites for the dissemina-
tion and contestation of ideas in the *bürgerliche Gesellschaft*,
the expanding civil society. As early as the 1790s, Humboldt
had wanted to participate in the public sphere. For him,
research always entailed communication; initially with his
fellow participants in Enlightenment salons, then with
his peers in the scientific community, and increasingly with
a public he addressed in his writings without knowing
them personally.

The care with which Humboldt nurtured his fame suggested more than a hint of vanity. From America, he arranged for the notes and letters he sent back to Europe to
be published so readers could keep up-to-date with his
spectacular journey. Following his return, Humboldt
worked on his public image for another reason. The travel
costs, disbursements, and other expenses incurred in his
life as a wandering private scholar had made a serious dent
in his private fortune. He therefore had an incentive to
make money from his books. This concern permeated his
correspondence with the publisher Johann Friedrich
Cotta, who signed up Humboldt as a promising writer in
1805. Cotta Verlag in Tübingen, later relocated to Stuttgart, was the premier publishing house in the German-
speaking world. It boasted Goethe, Schiller, and other
literary luminaries among its stable of authors.

Humboldt believed that his publications on his American journey would be "sellable" and attract interest even
from "unscientific readers," as he assured Cotta in 1805.[3]
This reflected a very modern way of thinking. Humboldt
was keenly aware that knowledge could not prevail through

its intrinsic worth alone. It had to be presented on the marketplace of public opinion and subjected to the laws of supply and demand. But this was easier said than done.

Humboldt's hunger for publicity could not be satiated until he had worked through the vast quantities of material he had brought back from America; they filled over thirty crates and numerous journal volumes. Together with Bonpland, he had collected over 6,000 species of plants.[4] Now he wished to supplement what he had learned on his travels with additional reading, collegial discussion, and new data. He initially spent almost nine months in Paris, which offered him the best working conditions. In a series of lectures held at the Institut National des Science et Arts, known simply as the Institut, Humboldt informed the French scientific elite about his American research. He forged close ties with the chemist and physicist Joseph Louis Gay-Lussac and the versatile Jean-Baptiste Biot. The two men had caused a stir shortly before Humboldt's arrival by ascending to an altitude of over 4,000 meters in a hydrogen balloon. Humboldt was especially interested in Gay-Lussac's experiments on the chemical composition of air and in Biot's geomagnetic expertise.

More than ever, Humboldt's science relied on collaboration with other experts. They were increasingly drawn from a younger generation that put a premium on professional specialization. Humboldt himself was no longer a young trailblazer but a fixed star in the transatlantic scientific firmament, notwithstanding Napoleon's dismissive treatment of him when they met for the first and only time in October 1804. The French emperor had little interest in

botany. Besides, from a political point of view, Humboldt
wanted to keep his options open. He was deferential to the
Prussian king, Frederick William III, who had taken the
throne in 1797. In this he was not without ulterior motives.
Despite his affinity for France, Humboldt depended on
Prussian support. Dismayed by the prospect of ongoing
warfare in Europe, he also drew closer to his remaining
relatives. In March 1805, he crossed the Alps to reunite in
Rome with his brother Wilhelm, who had been appointed
Prussian envoy to the Holy See three years prior.

In the 1790s, Humboldt had immersed himself in literature and art in Jena. Now he embraced the coterie of German classicists and artists living in Rome, sharing with
them his thoughts on the cultures and languages of pre-
Columbian America. An opportunity finally arose to scale
Mount Vesuvius, south of Naples. Humboldt applied his
practices from the Andes to the European volcanic world.
He was determined to clamber all the way to the top, paus-
ing along the way only to investigate the specific qualities
of the zones he passed through, and peer into the geological depths from the crater rim—and do so repeatedly, just
to be sure. The companionship of Gay-Lussac and Leopold von Buch, his trusted fellow student and geologist
from Freiberg, was particularly fruitful from a scientific
perspective. Both men joined Humboldt on his long journey back to the redrawn German territories.

In November 1805, after an absence of over nine years,
Humboldt saw Berlin again. Meanwhile, many more
people called it home, around 180,000, including military
personnel. Berlin had become the sixth-largest city in

Europe in terms of population.[5] The road network had improved significantly since Humboldt's childhood, and the city was on its way to becoming a cultural center in Germany. In 1800, the Singakademie, an urban choir founded only a few years prior, even gave the posthumous premiere of Mozart's *Requiem*. The Berlin of the early nineteenth century drew intellectuals and artists. The Romantic writer August Wilhelm Schlegel and the French woman of letters Madame Germaine de Staël moved to the city; others followed in their footsteps. The scientific scene, too, was expanding. The establishment of a new medical training facility, the Pépinière, was followed in 1799 by the Bauakademie, a school to train architects. The publishing industry boomed, with topics from the natural sciences, mathematics, and medicine ranking second in overall production behind light fiction. Berliners followed the general trend from intensive reading of a select few texts to extensive reading of a wider range of publications. Here, as elsewhere in Germany, people's reading ability made great strides.

Humboldt spent almost two years in his hometown. This was unusual for this hyperactive traveler and suggested a desire to settle down. The decision to opt for a slower pace proved rewarding. Humboldt could renew old acquaintances such as those with Henriette Herz, Rahel Levin, and the Mendelssohn family. His knowledge and fame began to pay off. Prussia granted Humboldt an annual pension of 2,500 talers, and the king appointed him as his chamberlain, officially placing him in the royal entourage. The Prussian Academy of Sciences had already made him an extraordinary member in 1800, when Humboldt ended his voyage on the Orinoco; this allowed him to give lectures from the academy's podium. But Humboldt was not entirely happy in Berlin. As so often in his life, he was becoming an insider and yet remained an outsider. Political developments once again contributed to this predicament.

In July 1806, sixteen German states seceded from the Holy Roman Empire, forming the Confederation of the Rhine. Not long after, the empire was formally dissolved. More was to come. The resistance of Prussia, in alliance with Saxony and Russia, led to a renewed outbreak of hostilities with France. Napoleon gained the upper hand. His troops defeated his enemies at the battles of Jena and Auerstedt. In October 1806, Napoleon entered Humboldt's Berlin. The French occupation lasted until December 1808. Locals were forced to billet French soldiers; the Humboldt family estate in Tegel was lucky to escape the spree of looting. Napoleon imposed substantial war contributions on Prussia and banned trade with Great Britain.

Humboldt was suffering. He had become too much a wanderer between different cultures and states to throw in his lot with the emerging nationalist movement in Germany. Militancy and punchy slogans had never been his style. At the same time, a "mood of melancholy" arose "from the dreadful state of my fatherland," as Humboldt summed up his dilemma.[6] In November 1807, acknowledging his role as a mediator, Prussia dispatched him to Paris on a diplomatic mission to reduce its compulsory payments to France. While Humboldt played only a bit part in the negotiations, the mission allowed him to establish a base in Paris again. With the Prussian king's blessing, he would make the French capital his home for almost the next two decades.

In the tense years between 1804 and 1807, and thus relatively soon after his return from America, Humboldt succeeded in documenting his geography of plants, in retrospect probably the most significant legacy of his scientific work. As it turned out, half a century of active scholarship still lay ahead of him. He had not yet fully analyzed his collection of botanical specimens. But Humboldt summoned his courage and published the French and German editions of his *Essay on the Geography of Plants* (*Ideen zu einer Geographie der Pflanzen*) almost simultaneously in 1805–7. His early exchange with the botanist Willdenow in Berlin, followed by the studies on subterranean flora in Freiberg, had provided him with a firm foundation. Humboldt's experiences in America had enhanced his long-standing interest in understanding plants in their natural habitats, through transregional comparison, and in their vertical distribution from caves to mountaintops.

Line by line, *Geography of Plants* illuminated these complex connections. Humboldt invited his readers to accompany him on his Andean ascents and on imaginary treks in the Himalayas, to Madagascar and Sumatra. He proposed a new typology of the "floral blanket" covering the earth. In essence, Humboldt suggested distinguishing plants that grew isolated and dispersed from those that were "socially united, like ants and bees."[7] What produced the world's abundance of plant life was thus not a primal form, as Goethe maintained, but the distinct interplay of a multitude of regional factors, periodically interrupted by natural catastrophes.

Humboldt carefully set out the conditions for this botanical profusion. Geological structures and soil composition were no less important than height above sea level, temperature, light intensity, and air quality. His reflections on this last factor owed much to the influence, since 1804, of Gay-Lussac. He also acknowledged the work of other pioneers such as Haenke, Jacquin, and the Swiss alpinist Saussure. Some readers may have recalled Humboldt's conclusions on the problem of galvanism from the late 1790s, when the author emphasized that in the "great chain of cause and effect" no phenomenon could be regarded in isolation: "equilibrium arises from the free play of dynamic forces."[8]

Humboldt's geography of plants allowed him to undertake a dual synthesis. First, he collated his empirical observations and made clear that there need be no categorical opposition between empiricism and natural-philosophical ideas. This was a conciliatory gesture toward Friedrich Wilhelm Schelling. His idealist philosophy of nature, which stood in high regard in these years, similarly valorized the concept of sensuous perception (*Anschauung*). Second, Humboldt adapted his research to other cultural practices of his time as well, just as he had attempted in 1790s Jena. He combined different representational forms and tailored them to the perceived needs of his reading public. The naturalist Humboldt recommended emulating landscape painting and artistically condensing scenes of nature in a

Totaleindruck, an "impression of the totality,"[9] a term previously used by Georg Forster. This aesthetic intention was made apparent in the lecture on the physiognomy of plants Humboldt gave in Berlin in January 1806. Excerpts from his travel reports, stirring passages, and references to research discussions went hand in hand with the call to "abstract from local phenomena" and use the imagination to create "vivid images" of nature.[10] Schiller's influence was again noticeable.

Schiller was no longer alive, so it was left to his fellow writer Goethe to sing Humboldt's praises. Goethe was much taken with the geography of plants, as Humboldt proudly noted. In fact, Humboldt had considerably distanced himself from his former mentor in the question of a botanical primal form and with his pivot to Plutonism. The work nonetheless turned into a homage to Goethe and to classical art that brought his Jena experiences into the new century. The German edition was prefaced by a dedicatory page designed in Rome in 1805 by the Danish sculptor and classicist Bertel Thorvaldsen. There Thorvaldsen had belonged to the circle around Wilhelm and Caroline von Humboldt. The vignette made direct reference to Goethe's elegy, "The Metamorphosis of Plants."

Humboldt used the term "painting of nature" (*Naturgemälde*) as a leitmotif in his geography of plants and throughout his later oeuvre. While he never precisely defined this powerful trope, he took it literally, too. The French and German editions featured a *tableau physique* or "Nature Painting of the Andes," first sketched in 1803

FIGURE 7. Excerpt from the "Tableau Physique des Andes" in the *Geography of Plants* (1807).

on the Pacific coast. It recalls a medieval triptych. Humboldt placed a colorful illustrated panel at the center, framed on both sides by vertical scales. The panel shows a cross-section of the South American continent at the altitude of Chimborazo. The aim was to combine precision with "painterly effect."[11] The proportions and other details were adapted and condensed for the purposes of artistic illustration. The third of the image on the left, facing the Pacific, shows the Andes viewed from the south. Humboldt positioned the smoking Cotopaxi volcano nearer to Chimborazo. The right-hand side of the image, turned to the Atlantic coast, presents an idealized cross-section of the mountain, as if Humboldt had cut it open and spread it flat. Humboldt entered the names of typical plant specimens he had collected here to indicate their distribution patterns. He added measurement data and scientific explanations on both sides.

We know today that Humboldt frequently made mistakes and that many entries on this tableau, drawing both on his predecessors' studies and the data he had collected in America, are incorrect.[12] Still, as an attempt to provide a "painting of nature," it ranks among Humboldt's most evocative works, attesting to the power of aesthetic imagination without conforming to the ideal of empirical objectivity.

Humboldt appealed to a wider readership in a collection of three essays, dedicated to his brother Wilhelm, which appeared in 1808 under the title *Ansichten der Natur* (*Views of Nature*). Here, too, Humboldt doffed his cap to classical German literature. His foreword culminated in four lines from *The Bride of Messina*, Schiller's tragedy.

Humboldt added his voice to its chorus: "On the mountains there is freedom!"[13] In two subsequent editions, the
author expanded the collection to seven pieces, including
"Rhodian Genius," his unconventional literary essay from
1795.

Views of Nature, whose title alluded to Georg Forster's
account of their shared journey in 1790, was Humboldt's
favorite child. His readers agreed. *Views* became his most
popular work and a landmark in the history of nature writing. In the original version, Humboldt first invited his
readers to join him in northern Venezuela, which he had
visited in early 1800. Next came his Berlin lecture on the
physiognomy of plants. The book concluded with depictions of the Orinoco rapids at Atures and Maipures, where
Humboldt and his companions had almost drowned.
Those interested in Humboldt's attempt at a dual synthesis
no longer needed to grapple with his *Essay on the Geography
of Plants*. *Views* familiarized them with Humboldt's
panoramic view of nature as a tableau in which nature's
attractions featured as a "picturesque spectacle."[14] The
essays emphasized the connection between empirical
research, sensuous perception, and aesthetic pleasure. The
fact that the opening pages immediately transported
readers to Africa and Asia showed the author's skill in
integrating comparative perspectives into his account.

Not coincidentally, Humboldt prefaced *Views* with the
same expression he would use almost forty years later in
introducing his magnum opus, *Cosmos*; he was now "deliv-
ering" his work "to the public."[15] That he did so "shyly," as
the author added in 1808, was a little coquettish. Humboldt

was not exactly afflicted with shyness when it came to sharing his ideas with the world. Yet there was more to this rhetoric than false modesty. Humboldt had recognized that in the nineteenth century, knowledge had become a commodity. The author could not predetermine his work's resonance and critical reception. Both would be decided and constantly reassessed on the marketplace of public opinion. Humboldt consequentially adhered to the maxim that "books . . . have no life without the public."[16] Together with the Cotta publishing house, he spent the rest of his life devising strategies to gain the audience's attention. He advertised his writings in periodicals and arranged for them to be translated from French into German and vice versa, later especially with a view to an anglophone readership. Through dedications, additions to the text, variant wordings, and footnotes, Humboldt adapted to the expectations of readers from different cultures. Goethe, for example, was omitted from the French edition of the geography of plants.

The two decades in which Humboldt had his home base in Paris, from 1807 to 1827, coincided with the middle phase of his life. Humboldt took up permanent residence in the French capital at the age of thirty-eight and left it at fifty-seven. He was generally in good health, troubled only by rheumatism in his right arm, which he blamed on the humid climate of the Orinoco. As a bachelor without family obligations, Humboldt remained a workhorse and got by on four hours' sleep a night.[17] We do not know whether any of his many male friendships involved physical intimacy. Humboldt closely guarded his privacy.

These years represented a new phase in Humboldt's life, despite all the continuities. Humboldt traveled, albeit more sporadically than before. Political events in Europe and requests from the Prussian government occasionally caught up with him. He continued to crave the independence of a scholar. Humboldt became the central node in a network of transnational contacts that radiated ever further outward. Paris was better suited to this than Berlin, and it offered better conditions as a scientific hub. For visitors from across the Atlantic, the French capital was the premier address in Europe, and the city lay closer to two other hubs, London and Madrid. It was in Paris that Humboldt first encountered Simón Bolívar from Caracas, the driving force behind the South American independence movements in the decades to come.

Humboldt was an integral part of the academic world of Paris, revered as a scientific titan despite being under police surveillance from 1807 due to alleged espionage. The Institut and the Société d'Arcueil, a new, elite scientific association, made him a member. From 1809 on, the close friendship with François Arago, a physicist and astronomer almost twenty years his junior, shaped Humboldt's life in Paris. Arago was a rising star in the French scientific community and later appointed director of the Paris Observatory. Arago introduced Humboldt to his pathbreaking research on the wavelength and polarization of light and on electromagnetic phenomena, in which Humboldt had long taken an interest.

All these circumstances left their mark on Humboldt's science. Before 1804 it had been thematically decentered, was based on firsthand research in many different locales in Europe and the Americas, and bore an experimental character, imbued with insecurities. After 1804, Humboldt attempted to do justice to the many kinds of information he had acquired and to address them separately, working industriously on several fronts. Now he set out to define his key ideas more precisely and make them the focus of his work.

Both *Essay on the Geography of Plants* and *Views of Nature* showed a newfound eagerness to draw together the threads. Humboldt sifted through episodes and local experiences in search of evidence for a "survey of nature at large."[18] In this respect, Humboldt's science was becoming more centered, as a repository of information, and increasingly shifting to the intellectual plane. Constant physical mobility gradually made way for a more settled existence in Paris, and later in Berlin. Humboldt tried to keep up-to-date with the latest scientific research through nonstop communication with specialists, who provided him with new information and confirmed or corrected his knowledge. This was an uphill battle, given the continuous expansion of knowledge in his time. Nonetheless, Humboldt retained his curiosity, his keen eye, and his ability to make surprising forays into new fields of knowledge. And he continued to bite off more than he could chew, tackling projects he could never possibly bring to completion.

These subtle shifts in Humboldt's science became apparent as he wrote up his American journey. This project came to dominate Humboldt's life during his years in

Paris. His decision to publish it in French gave rise to one of the many ironies that characterize his life and work. Unsurprisingly, German audiences focused on the works that came out in German, especially Humboldt's *Essay on the Geography of Plants*, *Views of Nature*, and later *Cosmos*. His earlier German publications were too complicated and had little appeal. Moreover, the French-language reports on the American journey and Humboldt's individual works on the subject, as well as his publications about Mexico and Cuba, had a hard time finding a German-speaking readership. They did so only indirectly, after many decades and in flawed translations.

From 1807 on, Humboldt published his travel narrative in installments in French. He gave it the general title *Voyage aux régions équinoxiales du nouveau continent* (Journey to the Equinoctial Regions of the New Continent). The work became so fiendishly complicated in its conception and variants that it still has the power to drive experts, librarians, and readers to despair. Different editions led to different volume counts (twenty-nine might be a reliable number)[19] and differences in content. The titles given to the volumes were not coherent. Publication dates are confusing, as most volumes were released in separate installments over many years. Not even Humboldt owned a complete edition. Changing publishers on the French side caused headaches. Production costs for the text and illustrations skyrocketed, making the work unaffordable for most. These costs ate into the author's private income, since he had to foot the bill for subsidies, collaborators, and missed contractual obligations. With

Voyage aux régions équinoxiales, Humboldt left behind a monumental work that also resembled a construction site of knowledge. In this respect, the American travelogue reflects Humboldt's science in its promises and difficulties.

It pained Humboldt and his public that his planned four-volume account of his actual travels—titled *Relation historique* in French, *Reise in die Aequinoctial-Gegenden* in German, and *Personal Narrative* in English—only began to appear in 1814 and remained incomplete. Humboldt could not extend it beyond his stay in the north of present-day Colombia in early 1801. His diaries, by contrast, covered the entire American journey. Between these original notes and the printed travel report, there were differences in style and content as well as textual displacements. The diaries were to remain Humboldt's most important source after 1804. With their help, he communicated a sense of direct experience to his readers, whether that of fending off constant plagues of mosquitoes or toiling up the slopes of the Andes. He coupled this authenticity with more general reflections on nature and society in the Spanish colonial empire. Humboldt was forthright in *Relation historique* in his condemnation of slavery. The same impression is conveyed in the English version that began to appear in 1814 and the German translations that were published from 1815.

The story of Humboldt's years in Paris could be told through the series of publications that the author produced in extraordinary profusion, despite all the obstacles in his path. Humboldt not only confirmed that he was a prolific writer but also revealed his inclination to work simultaneously on several manuscripts. One project was not enough, it seemed; he simply pushed himself to keep writing, on and on. Parallel to his work on *Relation historique*, in which he retraced the course of his American journey, Humboldt published from 1808 onward his *Essai politique sur le royaume de la Nouvelle-Espagne* (*Political Essay on the Kingdom of New Spain*), his work on Mexico. The French title was an understatement. Humboldt offered a comprehensive account of Mexico's landscape, economy, and history, which he had systematically researched from 1803 to 1804. The subtitle to the German edition—bizarrely long, not just from today's viewpoint—accurately reflects the scope of Humboldt's inquiry: *Containing investigations into the country's geography, its surface area, and its new political organization, its general physical make-up, the number and mores of its inhabitants, the progress of agriculture, manufactories, and trade, the proposed canal linking the Caribbean Sea with the great ocean, the military defense of the coasts, state revenue and the mass of precious metals that have flowed to the old continent, both east and west, since the discovery of America.*

With this complex work, based on firsthand observations and measurements as well as archival sources, Humboldt presented what was, besides the *Relation historique*, his most voluminous work on his American journey. It reflected the rise of modern-style political economy developing in Europe. Humboldt supplemented the text with cartographic material collated in a separate atlas.

The Mexican work was widely read as the first modern ethnogeographic survey of the Central American state, which seceded from Spain in 1810 and gained independence in 1821. Humboldt wanted to move beyond a mere accumulation of details to provide a "painting"[20] that would satisfy public expectations. He toyed with gestures of deference, such as the dedication to "His Catholic Majesty," Charles IV of Spain. After all, it was the Spanish colonial government that had made his journey possible. But he did not shy away from criticizing the "Europeans' cruelties."[21]

In preparing his American journey for publication, Humboldt invited colleagues to share their expertise in specific fields. German scientists were among those recruited as collaborators; he even convinced his old Berlin friend Willdenow to join him in Paris for a few months. Carl Sigismund Kunth, a nephew of his former tutor Johann Christian Kunth, replaced the less efficient Bonpland as his botanical consultant. Kunth was primarily responsible for writing the botanical section of the travel work. In addition, Humboldt relied on painters, graphic designers, engravers, and mapmakers to enhance the visual apparatus as a necessary complement to his texts.

A team of suitably qualified German and French artists also collaborated with Humboldt on *Vues des Cordillères*, completed in 1813. *Picturesque Views of the Cordilleras*, to use the English title, featured sixty-nine large-format illustrations. Accompanied by explanatory texts, they combined scenes of the landscapes traversed by Humboldt and Bonpland with sketches of the cultural "monuments" they had encountered along the way. Among the latter, Humboldt chose sketches of an Inca calendar, costumes of the

pre-Columbian population, and fragments of Aztec painting. Nature and art, he wrote Goethe when sending him the
first installment, were "closely intertwined" in his work.[22]

Humboldt was aware that *Picturesque Views* suffered
from a lack of logical structure and context, as did many of
his publications. In a disarming way, he never concealed
the danger he faced in placing observations "side by side"
rather than "penetrating the nature of things" to present
them "in their internal connections."[23] In *Picturesque Views*,
Humboldt walked a tightrope between wanting to make
continuous comparisons and avoiding the assumptions of
European cultural superiority that so often informed such
comparisons. Precisely in their eclecticism, this collection
of "monuments" mounted a powerful case for appreciating
American cultures, languages, and traditions in their own
right, and for European readers in particular.

Just as his publications on the American journey were
gaining momentum, politics once again got in the way.
The peace treaty with France concluded in Tilsit (today's
Sovetsk) in East Prussia in the summer of 1807 had given
Prussia a respite. The reorganization of German territories
under French pressure and Prussia's defeats had exposed
serious structural weaknesses. King Frederick William III
used the opportunity to entrust his most capable states-
men with instituting far-reaching reforms in society and
the state. Reforms were rolled out under the leadership of
Minister Freiherr vom und zum Stein, and from 1808
under his successor Hardenberg. They paved the way for
the transition from a feudal, estates-based order to a society based on bourgeois meritocracy. Prussia introduced

self-governance in towns and broke the guilds' monopoly on trade and industry. The kingdom modernized its bureaucracy and granted Jewish subjects much the same rights and duties as their fellow citizens, although with some exceptions. The army was reformed on the basis of compulsory male military service.

Stein and Hardenberg knew the Humboldt brothers well. In 1809, Hardenberg appointed Wilhelm to lead the directorate of education and culture in the Prussian Ministry of the Interior. He was tasked with reforming the education system and laying the groundwork for a new university. Due to shifts in territory over the preceding years, Prussia had been left with only three institutes of higher learning. Yet academic training was needed to keep pace with the increasingly rapid expansion of knowledge and to give the civil service and middle class the intellectual profile demanded by a modern, professionalized society. Having spent years advocating a neohumanist education, Wilhelm was not about to let the opportunity go to waste. The University of Berlin opened its doors to its first intake of students in the autumn of 1810. By that time, Wilhelm had resigned from his post for political and personal reasons.

When Wilhelm was sent to Vienna as Prussian envoy, Hardenberg offered his directorship to Alexander, who had worked for him in Franconia in the 1790s. Humboldt declined. Six years later, when Hardenberg offered him the position of Prussian ambassador in Paris, Humboldt again turned him down. His scientific work came first. Besides, Humboldt was contemplating an extended voyage to Central Asia, India, and possibly even the Philippines. Despite several attempts to launch the trip in the following years, the plan foundered on resistance from England, which dominated the southern regions of Central and Asia, and the English East India Company, which controlled trade with India.

During his time in Paris, political events continued to make their presence felt in Humboldt's life. Napoleon had been on the retreat in Europe since the autumn of 1812. With significant Prussian involvement, the coalition of forces arrayed against him took the fight into French territory, occupying Paris in early 1814. Wilhelm, unabashed in his support of the German cause, was there with them. It was a cause his brother, *citoyen Humboldt*, "Citizen Humboldt," could not embrace, souring relations between the two. Wilhelm was not amused. Once again, Alexander found himself caught between the fronts. King Frederick William III sought out his company in Paris even as Humboldt kept his distance from anti-French conservatives from Prussia and Austria. He helped save the Museum of Natural History in Paris from looting and forced contributions to the victors. Humboldt came under fire from nationalist circles in Germany when he successfully lobbied to have art treasures stolen from German territories by Napoleonic troops remain in Paris.

Over the following years, Frederick William III occasionally claimed Humboldt as a consultant and travel companion, inviting him along to London in 1814, to Aachen in 1818, and to Verona in 1822, where the king took part in the congress of the conservative European powers.

For the Austrian statesman Clemens von Metternich, the epitome of the forces of reaction in Europe, Humboldt remained a "political maverick."[24] The description was apt, for Humboldt was not known for his subservience to his political masters. Observers had for long remarked on Humboldt's criticism of colonial exploitation in South America and his closeness to European liberals.

As was his custom, Humboldt deftly combined these official travels with private scientific excursions; in 1825, he added a short trip to the Bretagne. In Italy, Humboldt even managed to revisit his beloved Vesuvius. London held an obvious attraction for him as a center of global commerce in people, goods, and knowledge; he paid the city another two visits. An additional incentive was that Wilhelm resided there since 1817 as Prussian ambassador. Humboldt continued trying to obtain permission for a journey to India. In London he became reacquainted with Joseph Banks, whom he had first met when traveling with Forster in 1790, and with the German-born astronomer William Herschel, whose enormous telescope he admired in 1817.

On his return journey from his final stay in Italy in 1823, Humboldt visited Berlin and Tegel for the first time in sixteen years. He was slowly circling back to the scenes of his childhood and youth. But several scientific projects in Paris still needed to be concluded. In 1804, Humboldt and Biot had used lines of equal magnetic field intensity in an article.[25] Thirteen years later, Humboldt put forward a similar idea in a French-language publication that took comparative climate research to new heights. Humboldt

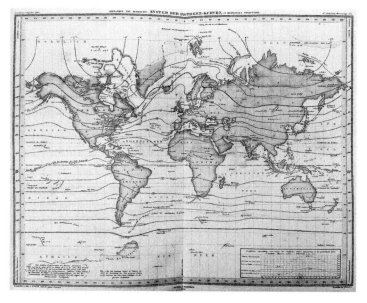

FIGURE 8. "Alexander von Humboldt's System of Isothermal Curves" (1845). Created by Heinrich Berghaus (1849).

recommended connecting all the places on earth with the same average temperature using so-called "isothermal lines" (*lignes isothermes* or *lignes d'égale chaleur*).[26] Although this idea built on eighteenth-century precedents, Humboldt adapted it in an innovative way to demonstrate the theorem posited in his work on Mexico: "No point is unrelated to any other."[27] Humboldt's isotherms came to rank among his most influential achievements; in this respect, they complemented his geography of plants. Both made it possible to measure the world in ways that transcended political borders and to demarcate ecological zones spanning entire continents and oceans.

Besides this proposal and volumes on specialist topics, Humboldt's *Essai politique sur l'île de Cuba* (*Political Essay on the Island of Cuba*) takes pride of place among the works he wrote near the end of his time in Paris. His two stays on Cuba, in the winter of 1800–1801 and the spring of 1804, had familiarized him with the island. *Political Essay* was a composite of previously written texts, again resulting in a complex textual history. In attempting to present a comprehensive overview of the geographical, economic, and sociological characteristics of an entire region, it resembled its predecessor volume on Mexico. The detailed treatise emphasized the problems of economic monocultures, especially sugarcane plantations, and criticized the exploitation of human and natural resources. Humboldt again drew attention to the line of historical development running from the "tyranny" of the Spanish conquistadores to the greed of modern-day enslavers.[28]

As so often before, however, he distanced himself from revolutionary violence and a reign of terror, such as that undergone by France between 1793 and 1794, as radical cures for the blight of slavery. Humboldt had witnessed the devastating effects of national wars on Napoleonic Europe, and they filled him with dread. He remained committed to liberal reforms, advocating the gradual transformation of a deeply hierarchical, economically exploitative, and inefficient colonial society. Here, too, Humboldt remained entrenched in middle-class values.

Through his travels and publications, and not simply as a critic of slavery, Humboldt played a key role in deepening European interest in South America. Conversely, he provided legitimation for those on the other side of the Atlantic who, with Bolívar, were beginning to call for independence from the Spanish colonial empire. This was not without a certain irony. Humboldt had left South America years before the Spanish colonial empire splintered into independent nation-states. He played no direct role in the volley of independence declarations that in the years between 1811 and 1825 led to the foundation of Paraguay, Argentina, Chile, the Republic of Greater Colombia (from which Colombia, Venezuela, and Ecuador later emerged), Peru, Mexico, and Bolivia. Yet he welcomed this development even where he continued to see the new elites critically. Humboldt became a positive point of reference for many South Americans in the nation-building process and continued to be venerated on their continent.

While calling Humboldt the 'second discoverer of America' is a myth born out of a colonial mindset, the label 'father of Latin American independence' is an equally ahistorical, heroizing trope that served those seeking a European source of legitimacy after colonialism. On both sides of the Atlantic, Humboldt's contribution to understanding South America was far more nuanced. His account of nature and society, for all that it remained part of an open-minded, "*inclusive* Eurocentrism,"[29] fostered mutual understanding between Europe and America on a scientific basis.

During his Paris years, Humboldt built up a network of contacts in English-speaking North America to complement his existing connections in Latin America. This made him one of the first Atlanticists of the modern world. Humboldt corresponded with Thomas Jefferson, Secretary of State James Madison, and Treasury Secretary Albert Gallatin, all of whom he had gotten to know in the United States. An especially lively exchange ensued with the Swiss-born Gallatin, who had a keen interest in pre-Columbian ethnology. The circle expanded to include visiting American diplomats, scientists, and writers such as George Ticknor, who had studied in Göttingen for two years and taught Romance languages at Harvard. These exchanges were always mutually advantageous. Humboldt gave his American guests access to his vast network of European contacts. In turn, they provided him with references and information from the natural sciences. Many of them helped publicize Humboldt's writings in the United States.

By the mid-1820s, Humboldt was renowned and even revered for his scientific exploits and far-flung travels. He was lionized in Paris, in Europe, and across the Atlantic. The political climate in France was deteriorating, however. Archconservatives were gaining ever more influence under King Charles X, on the French throne since 1824. Several achievements of the French Revolution had been wound back. Prussia, too, became one of the protagonists in the period of restoration and reaction that set in after the Vienna Congress of 1814–15. Still, the Prussian king esteemed Humboldt and was keen to draw him into his orbit. This was not the worst scenario for Humboldt, given the precarious state of his finances. His brother Wilhelm had resigned from state service in 1819, retiring to the old family estate at Schloss Tegel to devote himself to studies in linguistics; he was soon beset by various ailments. Thanks in part to his initiatives, Berlin had become a more interesting place to do science. Humboldt thus had good reasons for coming home.

6

I Love What I Comprehend

IN CHANGING WORLDS,
1827–1840

WHEN HUMBOLDT moved from Paris to Berlin in May 1827, at the age of fifty-seven, both the city itself and Prussia were transforming in profound ways. Prussia's international standing, too, had changed dramatically. Following Napoleon's abdication in April 1814, the European powers had met at the Congress of Vienna to redraw the political map of the continent. Wilhelm von Humboldt had been involved in negotiations as an envoy. He had worked tirelessly to draft and redraft diplomatic proposals, and he shared the general "nervousness" of all participants.[1] Meanwhile, Alexander did his best to avoid taking a clear political stand. Europe's new order was decided on even before Napoleon, returned from exile in March 1815, was defeated one last time at Waterloo. Prussia gained substantial territories in the west, which were amalgamated into the provinces of the Rhine and

Westphalia. With that, the kingdom's borders moved much nearer to Paris.

In the new German Confederation, a grouping of initially thirty-eight territories that was in some ways the successor to the Holy Roman Empire, Prussia was rivaled only by Austria as the dominant power. Both joined Russia in the conservative Holy Alliance, aimed at providing the continent with political stability under conservative, Christian auspices. Like other German states, Prussia adopted the Carlsbad Decrees of 1819, directed against liberal and national ideas. King Frederick William III and his ministers tightened censorship, curbed freedom of expression, and suppressed dissent, including at universities. The French government thus did not stand alone in its eagerness to appease the forces of reaction, much to Humboldt's frustration. Having sought a more conducive climate in Berlin, he now had cause to poke fun at the pall of conservatism that had descended on his homeland.[2]

Yet the period leading up to the 1848 revolution was not defined solely by conservative restoration or a mood of political resignation. The middle classes did not simply retreat to the castles in the cloud of Romanticism, nor did they barricade themselves within the tastefully decorated interiors later given the name *Biedermeier*. They continued to strive for economic success and participation in the public sphere. In addition, the population increased significantly and social pressure built up from below due to the rising number of industrial workers. All these factors exacerbated the contrast between state-sanctioned conservatism and the dynamism of civil society, further complicated by the growing claims of a nascent proletariat. Scientific life was not exempt from this tension, which Humboldt experienced both in Prussia and in the new worlds he explored between 1827 and 1840.

Berlin's population doubled to over 400,000 inhabitants between around 1820 and 1848, primarily due to internal migration from the countryside. Berlin now played in the premier league of Europe's metropolises.[3] The city expanded and became more densely built. New industries settled in the urban area as technological progress transformed everyday life. Humboldt witnessed the advent of gas lighting and early forms of public transportation. From 1832, Berlin became linked by telegraphic lines with other regions. In October 1838, the first train steamed into Potsdam station, shortening a route that Humboldt had previously taken on horseback or by carriage. Demand for new railways helped stimulate industrialization. Cultural developments and education policy soon followed suit. Imposing neoclassical buildings were erected in Berlin to designs by architect Karl Friedrich Schinkel. Schinkel was even commissioned by Wilhelm von Humboldt to renovate Schloss Tegel. The Prussian education minister Altenstein modernized the school system and sought to attract or retain high-profile scientists. By 1834, the number of university students had risen to over 2,000. By the end of the decade, they were being taught by almost 100 professors—all men.[4]

Against this backdrop, and just six months after his return, Humboldt sent a signal that drew widespread attention. In November 1827, he gave the first in what would grow to a series of sixty-two lectures at Berlin University. As a member of the Prussian Academy of Sciences, Humboldt was entitled to lecture despite not holding a university chair. Due to unprecedented public interest, a parallel lecture series for a bigger audience was organized at the concert hall of the Berlin Singakademie, located near the grand boulevard Unter den Linden; today, the building houses the Maxim Gorki Theater. Carl Friedrich Zelter, the Singakademie's director and a friend of Goethe, was receptive to new tunes. The Singakademie provided the now world-famous Humboldt with a splendidly resonant auditorium.[5]

The sixteen lectures at the Singakademie were documented by Humboldt and others in notes and transcriptions. Although the lectures made considerable demands on their audience, Humboldt delivered them so skillfully that listeners were engrossed and even enthralled. Not without pride, he noted that the "liveliness" of his language had "some effect on the public."[6] Humboldt became the talk of the town, drawing royalty, officials, military personnel, ordinary citizens, and, according to contemporary reports, an unusually large number of women to his talks. Humboldt estimated that he addressed an audience of around 1,500 at the Singakademie each week.

The Singakademie lectures broadened the geographical and intellectual horizons of Berlin society and demonstrated the orator's far-reaching approach to nature and the world. Humboldt painted his scientific worldview on a grand canvas that defied the myopic views represented by the politics of restoration. He demonstrated the limitless interrelations of all natural spaces, which form a single ecosystem (as we would call it today) together with the humans who intervene in them. His lectures unfurled an "image of the totality of nature" (*Bild eines Natur-Ganzen*),[7] which he had been developing since his geography of plants and *Views of Nature*. Humboldt began with the planetary system and geological structures of the earth's crust before turning to meteorology and the geographical distribution of animals. Having touched on the history of humankind, including a condemnation of slavery, he addressed questions of electricity and magnetism before returning to astronomy. The lectures ended with an appeal for the sciences to remain open to aesthetic descriptions of nature.

Humboldt made clear that he was moving away from a purely geographical account, anchored in the *physique du monde* of the eighteenth century, toward a "physical description of the world"[8] that he would soon dignify with the time-honored title, 'cosmos.' He made various statements on how his university lectures and the Cosmos lectures of 1827–28, as they came to be known, contributed to the genesis of his eponymous late work. From 1803, when he began work on *Naturgemälde der Tropen*, his "painting of nature in the tropics," Humboldt made repeated trial runs at the Cosmos project. After 1827, he reflected more methodically on his "extravagant idea of describing in one and the same work the whole material world . . . from the nebular stars to the mosses on the granite rocks—and to make this work instructive to the mind, and at the same time attractive, by its vivid language."[9] This

is how he put it in 1834 in a letter to his friend, the democratic writer Karl August Varnhagen von Ense.

The Cosmos lectures marked a new phase in the history of public awareness of the natural sciences in Germany and elsewhere, giving these a prominence in civil
culture they had previously lacked. Contemporaries saw
them, together with *Views of Nature*, as heralding the birth
of a 'popular science' movement that would redefine the
relationship between professional expertise and the educational needs of society at large.[10] Over the long term,
the Cosmos lectures played a key role in integrating the
empirical natural sciences into a nineteenth-century culture dominated by literature and the arts. They helped free
scientific education from the stigma of the second-rate at
a time when neohumanists were still fending off the advance of the natural sciences in schools and universities.
Popularization made a difference.

Humboldt was keenly aware of the increasing pluraliza-
tion of the public sphere. He sensed that the homoge-
neous, ideal public of his youth was fracturing into ever
more diverse publics, each sustained by a different social
class and with its own demands for knowledge. These
called for new communication strategies in words and
images, in lectures and through educational institutions.
Associations and societies provided an important basis for
this work. The number of natural-historical associations
established in Germany, as well as their overall member-
ship, rose appreciably from 1830.[11] In these associations,
citizens—still overwhelmingly male—seized the opportunity to fraternize with like-minded peers. They drew

from their local surroundings and immersed themselves in scientific research. Humboldt encouraged such initiatives. Scientific education could only gain in status if it was transmitted through active community engagement, emerging from the midst of society rather than being imposed from above. In his Cosmos lectures, he therefore cited botanical gardens as exemplary sites of empirical instruction. Popularization was no one-way street. It meant an interactive process, based on the participation of ever-growing parts of society.[12]

The role of scientific associations as multipliers and disseminators of knowledge was on Humboldt's mind when he joined with Martin Hinrich Lichtenstein, the director of the Zoological Museum, in September 1828 to organize the annual conference of the Society of German Natural Scientists and Physicians (Gesellschaft Deutscher Naturforscher und Ärzte) in Berlin. Founded in 1822, the society provided a forum for convivial exchange between naturalists, scientific amateurs no less than university professors. They would convene each year at a different location.

A record number of guests, close to five hundred, gathered from all over Germany for the meeting in Berlin. Humboldt's contribution was crucial in making the conference a high point in the history of scientific events in Berlin. He was a genial host, even inviting one distinguished guest, the brilliant mathematician Carl Friedrich Gauss, to stay in his apartment. For the society's subsequent development, it was significant that Humboldt recommended holding panels on specialized topics alongside plenary sessions. In doing so, he made clear that general and expert discussions represented two complementary options in the broad spectrum of contemporary knowledge.

Humboldt's insight into how both knowledge and the public sphere were pluralizing prompted him to introduce two further innovations in Berlin. He lobbied to have the city's existing observatory upgraded to meet the highest technological standards since it was urgently needed for astronomical research. Humboldt also pressed for a new observatory that would open its doors to the public two evenings a month; construction was completed in 1835. And he threw his weight behind the proposal to set up a publicly funded zoological garden, modeled on the London Zoo. It welcomed its first visitors in 1844.

Still, Humboldt had not settled down entirely in Berlin. Travels continued to punctuate his life. The most important was the fulfillment of a long-held dream and began a year after the Cosmos lectures had ended. In April 1829, he set out on a journey that would take him east of the Ural Mountains for the first time in his life. During his Paris years, Humboldt's plans to visit the Himalayas, Tibet, and India had come to nothing. In 1827, correspondence with the Russian minister of finance, Georg von Cancrin, led to an invitation for him to visit Russia. Born in Hesse, Cancrin was a political reformer of a mercantilist bent. He found himself confronted with a host of economic challenges in the expanding Russian Empire, with its fossilized social structures and serf system. Cancrin was especially interested in modernizing the mining and salt industries as well as in currency reform. With that in mind, he hoped that Humboldt would compile a detailed

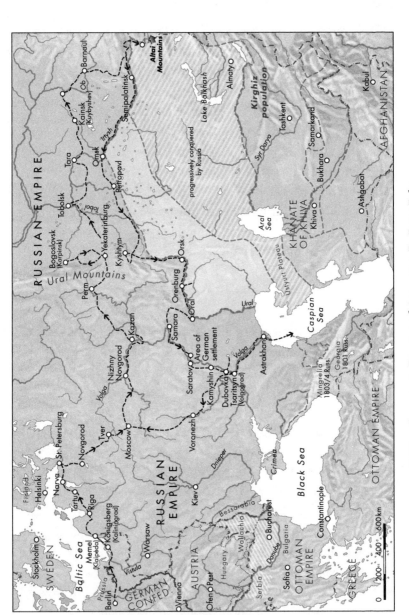

MAP 4. Journey to Russia and Asia, 1829. By Peter Palm.

report on platinum deposits in the Ural region. Russia was toying with the idea of issuing a platinum-based coinage.

Humboldt must have felt a sense of déjà vu. More than thirty years earlier, as a young official in the Prussian mining service, he had occupied himself intensively with similar themes. The Spanish and U.S. governments had likewise anticipated that his findings from the American journey would lead to improvements in raw materials extraction in their territories. It is therefore unsurprising that Humboldt's voyage to Russia followed the pattern set by his American journey. He traveled under government protection, this time even under Cossack guard. Such tutelage could prove onerous, especially as Humboldt was obliged to attend numerous official events. He inspected mines and tirelessly gathered data using the same, only slightly improved scientific equipment. Along with his Berlin manservant, Johann Seifert, Humboldt also again traveled with an entourage of scientific advisors. He was accompanied by the zoologist Christian Gottfried Ehrenberg, a veteran of Near East travel, and the mineralogist Gustav Rose, as well as a shifting escort of local guides.

The tour of Russia was more compressed, however, lasting only eight months, from April to December 1829. The Russian Emperor Nicholas I provided the necessary funds. By May, the travel party had arrived in St. Petersburg, the capital of the Russian Empire. En route, Humboldt first encountered the Baltic region and its scientific center, the Estonian town of Dorpat, today's Tartu. Since the eighteenth century, Russia had been drawing on the expertise of German-speaking Balts and other naturalists from western and northern Europe to lay the scientific basis for its eastward expansion and to chart its new territories more accurately. Humboldt was, thus again, not the first researcher to have traveled these parts of the world nor was he operating in a terra incognita.

Proceeding via Moscow, Humboldt and his fellow travelers reached Yekaterinburg in mid-June 1829. There, east of the Urals, he had arrived "in Asia," as he wrote his brother.[13] Geographical mobility and a comparative perspective sometimes brought together what lay far apart; Humboldt felt reminded of the heathland near Tegel. Continuing east, he followed the route recommended by his sponsor as far as Tobolsk. Humboldt inspected mines, took samples of gold, platinum, and topaz, and made geomagnetic, meteorological, and astronomical measurements. The group was supposed to turn back at this point, but Humboldt presented Cancrin with a fait accompli and led the group further east. They passed through Kainsk (Kuybyshev) on the Om and Barnaul on the Ob until they reached the western foothills of the Altai Mountains. Humboldt could not resist inspecting the mountains, mine shafts, and mineral deposits there.

In the source region of the Irtysh River, Humboldt came across a Chinese border post. The Germans and Russians in his travel party now found themselves face-to-face with Chinese and Mongols on the other side. Transposed to a political map of today, Humboldt was in far eastern Kazakhstan, not far from the border to China and Mongolia, which belonged at the time to the Qing Dynasty. With that, he had traversed well over half of the earth's circumference since 1799, a distance stretching from the western border of China via Europe and across the Atlantic to Mexico. Unlike Georg Forster and the German poet-naturalist Adelbert von Chamisso, another acquaintance, Humboldt never ventured into the Asia-Pacific, leaving aside a voyage up the west coast of America in 1802–3. This was not for lack of interest; even in his Paris years, Humboldt had been fascinated by Asia.

The Russian-Chinese border was the furthest east he would ever set foot. From there, Humboldt traveled back to the Urals, where he celebrated his sixtieth birthday. In Orenburg he again prevailed over his hosts. The group headed southwest to the Caspian Sea. On the way, it passed through territories settled the previous century by Germans at the initiative of Empress Catherine II. For the Russian Empire, access to the Caspian Sea was important for economic and strategic reasons. Part of the trade from Persia, the Near East, and South Asia ran along the Volga, which drains into the Caspian at Astrakhan. Humboldt was struck by the ethnic and cultural diversity he encountered at this crossroads between Europe and Asia. He met Armenians, Tatars, and Georgians; Indians, Persians, and Turkmens; Kirghiz and Kalmucks, descendants of a western Mongolian people. As on his American journey, Humboldt seized the opportunity to compile ethnological observations and ethnolinguistic documents.

On the Caspian Sea, unlike on the Orinoco and in the Caribbean, Humboldt could avail himself of a newfangled technology: steamships. The party proceeded up the Volga, then via Moscow to St. Petersburg, where Humboldt was greeted as an all-conquering hero and showered with gifts and honors. In a widely reported speech to the Russian Academy of Sciences, Humboldt recommended that the empire implement one of his most important recommendations for combining scientific research, agricultural improvements, and primary resource extraction.[14] He urged Russia to set up a far-flung network of observation stations for systematically gathering and analyzing geomagnetic and climate data. This would have the additional benefit of allowing the isothermal lines covering the Eurasian continent to be mapped with greater accuracy.

With this vision, Humboldt's science provided a model that morphed into what American historian of science Susan Faye Cannon has called 'Humboldtian Science.'[15] Contemporaries concerned with mapping, measuring, and scientifically exploring large territories (and thus controlling them, too) promoted respective research agendas. They used new technologies, especially the telegraph, to collect and transmit data, aiming at the continuous comparison of observations on a transcontinental scale. These efforts were meant to rely on state-run institutions acting in concert with individual scientists, whose work was often externally funded.

'Humboldtian Science' in that sense drew on Humboldt and his ideal of systematic measurements and quantification but jettisoned his emphasis on the aesthetic. Russia took up Humboldt's Petersburg proposal. Several years later, in 1836, Humboldt addressed the Royal Society in London and recommended creating a network of geomagnetic stations in the British colonial empire. The British government agreed. From 1839, Great Britain began setting up observation posts from Toronto to Cape Town, an initiative described at the time as a 'magnetic crusade.'

Having covered around 19,000 kilometers, Humboldt's Russian journey ended in December 1829 with his arrival in Berlin. By the end, the carriages were so crammed full of collected objects that they resembled a natural history cabinet. Humboldt was content and felt inspired to resume work on his magnum opus. The man was "like a bubbling pot," Zelter, the Singakademie's director, reported to Goethe from Berlin.[16] Far exceeding the parameters set by Cancrin, the journey had allowed Humboldt to broaden his views on comparative geography and climatology. The gains for the Russian government were more modest, especially as it took no pleasure in hearing Humboldt rail against the system of serfdom.[17]

As after his far longer trip to America, it was not easy for Humboldt to publish the results of the journey in a compact format; as so often, he needed helpers. Humboldt again presented a miscellany of text types. Based on lectures held from October 1830 in Paris, his *Fragments on the Geology and Climatology of Asia* appeared in French in 1831 and in German translation soon after. They concentrated on observations about mountain formations and the geomagnetic-climatic conditions of the enormous continent. Humboldt's guiding ideas shimmered through his dense scientific explanations. He refined his volcanic interpretation of the origin of mountains, speaking of the "mountain systems" of the Himalaya and Altai, which had arisen from the interplay between deep geological movements and changes in the earth's atmosphere. He grasped these processes as part of a comprehensive "physics of the earth."[18] Above all, Humboldt was finding his way to a "theoretical climatology." The irregular patterning of isotherms, their "diffractions," in Humboldt's words, gave him the tools he needed.[19] He emphasized that climate regions—and hence also the earth's flora and fauna—were distributed unevenly across shifting zones, rather than being grouped symmetrically along parallels north and south of the equator. In turn, these zones arose from interactions between regionally varied atmospheric and geographical factors.

Ever since his return from America in 1804 and his fruitless attempts to launch an expedition to India and the Himalayas, Humboldt had tried to learn more about Asian geography and languages. An important interlocutor in this regard was Carl Ritter, who had taught at Berlin University since 1820. Ritter was largely responsible for establishing geography as a scientific discipline in Germany. He produced an impressive body of work that, like Humboldt's, combined geography with ethnography. Furthermore, Humboldt profited from the expertise of Sinologist Heinrich Julius Klaproth, who had traveled the mountains to the east and south of the Russian Empire and resided principally in Paris. Klaproth made significant contributions to *Fragments*. Following his death in 1835, Humboldt stayed in contact with Orientalists. He learned a great deal from the Bavarian botanist, ethnologist, and Japanologist Philipp Franz von Siebold. In the 1850s, he took a keen interest in the expedition that the brothers

Hermann, Adolf, and Robert Schlagintweit mounted to India and the Himalayas.

Having returned from Russia, Humboldt initially invited his companion Gustav Rose to write a general narrative of their journey, giving him access to his travel diaries to that end. The two volumes of Rose's *Journey to the Urals, the Altai, and the Caspian Sea* came out in 1837 and 1842, respectively. Yet Humboldt soon began super-imposing new layers of text over those already written, an old habit going back to the 1790s. Under the heading *Asie centrale*, he worked from the late 1830s on a more compre-hensive account of what he had learned in Russia. The first two volumes, dedicated to the czar, appeared in 1843. What Humboldt called the "most industrious"[20] of all his works was not as innovative as he made out. It recycled a great deal of previously published material, supplemented by numerous specialist contributions. *Asie centrale* had little impact in Germany, for all that Humboldt explicitly presented it as a bridge between his works on America and the forthcoming *Cosmos*. It lacked the popular touch that Humboldt had promised himself from his "painting of nature."[21]

Those of Humboldt's publications that dated back to the 1830s were mainly written in Paris, even though he had relocated to Berlin in 1827. Humboldt's arrangement with the Prussian court gave him leave to spend up to four months each year in the French capital, where he found ideal conditions for researching and writing. For his part, King Frederick William III knew how useful Humboldt's numerous contacts in Paris were for him and Prussia.

Salons of all political stripes accepted the cosmopolitan Humboldt. This was even more important after the July Revolution of 1830 had forced the abdication of the conservative Charles X. Revolutionary impulses continued to agitate Europe in 1830 from Warsaw to Italy. In Brussels, citizens revolted against the rule of the Kingdom of the United Netherlands, which had been created by the Congress of Vienna, and called for Belgian independence. There was even some unrest in Berlin in September. The international situation was tense. Russia supported the Greeks in their struggle for independence from the Ottoman Empire, while France had sided with Egypt in their own rebellion against Ottoman rule in the so-called Oriental Crisis.

When the liberal Duke Louis Philippe of Orléans replaced Charles X as 'citizen king,' Frederick William III accepted the regime change over the objections of Prussian conservatives. Humboldt soon gained access to the court of the new French king. For Frederick William, this was a welcome development; he was convinced that Humboldt was politically harmless and lacking in personal ambition. His second-born son, the future Prussian king and German emperor, William, thought otherwise. Like many in Prussia's Foreign Office, he was dismayed to see his father repeatedly send Humboldt on diplomatic missions in Paris. From 1835 to 1847, Humboldt penned over fifty reports for the Prussian king summarizing France's domestic and international situation.[22] He spent a combined total of almost four years in the French capital.

Franco-German cooperation in cultural and especially scientific matters had been a constant in Humboldt's life. Since the years of the French Revolution, he knew only too well that political movements on both sides of the Rhine tended to view bilateral differences through more than just a strategic and military lens. These now crystallized into opposed political ideologies, mentalities, and national identities. These processes of nationalization on both sides intensified even before a German nation-state was proclaimed—in Versailles, provocatively enough—following the Franco-Prussian War in 1871. Humboldt was spared having to see this. He himself and his activities defied categorization in such terms. Humboldt was the catalyst for a politically fraught but lasting entanglement of German and French culture. He resisted all attempts to fix ideological and political boundaries. To that extent, his declaration of allegiance to France as his "second fatherland" (*seconde patrie*)[23] was a rhetorical gesture that reflected his pragmatic view of the world.

From 1830, Humboldt expanded his dense network of French correspondents. All his contacts benefited from his pragmatism and his role as middleman between Germany and France. He opened doors to French science for the chemist Justus Liebig and the zoologist Ehrenberg. The Egyptologist Richard Lepsius and architect Jakob Ignaz Hittorff, a longtime resident of Paris, likewise profited from his influence. Humboldt corresponded with the unorthodox German writer Heinrich Heine and met him several times in Paris. He promoted the Parisian reception of the architect Schinkel, and he lent his support to numerous French scientists. He also cultivated leading political representatives of the so-called July Monarchy, including the liberal writer François Guizot and the historian Adolphe Thiers.

The links forged by Humboldt brought together diverse social groups to form a microcosm of nineteenth-century intellectual life. His status as a free-floating intellectual made it easier for him to mix in circles ordinarily separated by social conventions and nominal borders. Topics of discussion ranged from the weighty to the mundane. They encompassed countless scientific details, career advancement opportunities for younger researchers, and avenues for publication and collaboration. Humboldt also liked to exchange academic and political gossip, especially when he felt safe to indulge in irony and sarcasm. In July 1837, commenting on a meeting between Frederick William III and the Austrian chancellor Metternich in the Bohemian town of Teplitz, today's Teplice, Humboldt scoffed at the ritualized spectacle of "'world-elephants' locking trunks" without any intention of altering the political status quo.[24]

Working so industriously in alternating and ever-changing worlds—in Berlin and pre-revolutionary Prussia, in Russia and Paris, and on occasional journeys with his king—while conducting a vast correspondence and mediating indefatigably between cultures would have exceeded the capacities of most contemporaries. Not so Humboldt. He even found time to supplement his unfinished American travel work with a section that had been close to his heart for decades. Between 1834 and 1839,

Humboldt published *Critical Investigations into the Historical Development of Geographical Knowledge Respecting the New World* in a German edition; a French version had earlier appeared as *Examen critique de l'histoire . . . .* Humboldt drew on a wealth of historical literature, including ancient writings, to give a detailed account of geographical exploration in the Americas from Columbus to the late sixteenth century. His particular interest lay in the text sources and cartographic material that the European 'discoverers' had used to gather information and produce a topography of America according to Western standards.

Like so many of Humboldt's projects, this one took on a life of its own. The text was originally planned as a commentary on maps, elevation profiles, and other drawings that Humboldt had started during his American journey and first compiled in 1814 in *Atlas géographique et physique*. His visits to Parisian libraries over many years allowed him to accumulate ever more reading materials and resources. The text grew to such proportions that it was published separately but, just like his travel account, remained incomplete.

Critical Investigations is among Humboldt's least-read books, despite its intriguing contribution to the history of science. Humboldt added another layer of self-reflection to the empirical contemplation of nature he had been developing so vigorously in his scientific works. He historicized what had often appeared as a side effect of European expansion into a supposedly New World. He connected changes in how the natural sciences perceived the world with one of the key intellectual trends of his time—the turn to history as a fundamental perspective for explaining the world beyond philosophical systems, considering each development in its specific context. *Critical Investigations* reflected the new historicism of the nineteenth century. By meticulously evaluating everything he read and "advancing to the sources,"[25] Humboldt subscribed to the methodology of critical history and source analysis, which Leopold von Ranke, the prominent Berlin historian, declared the prerequisite for objective knowledge and central to the emerging field of history.

Humboldt accepted the principles of historical understanding that the German historian Johann Gustav Droysen was developing at the same time. This receptiveness to the human sciences was instilled in Humboldt's life early on. During his studies in Göttingen in 1789–90, the classical scholar Christian Gottlob Heyne had helped him refine his historical interests. His brother Wilhelm had opened his eyes to the significance of linguistic comparisons since the 1790s. He had frequented archives in Franconia and Mexico. Since returning to Berlin in 1827, Humboldt had been in regular contact with the classical scholar August Böckh, who taught at the University of Berlin. In the 1830s, he had even attended Böckh's lectures and those of his French colleagues in Paris.

By the end of the decade, Humboldt had thus arrived at an impressive overview of contemporary knowledge. Now in his seventies, he dropped half-ironic, half-serious hints to friends and relatives that he had no wish to become a "fossil."[26] He still made do with only a few hours' sleep and worked late into the night. Yet writing had become more

onerous and turned into "a kind of toothache."[27] The period following his return from Paris in 1827 confronted
Humboldt with all too many reminders that life was finite.
The generation into which he had been born was fading
into history. Thomas Jefferson and Simón Bolívar, who
represented the independence of the new republics in
North and South America, died in 1826 and 1830, respectively. His old teachers Kunth and Blumenbach breathed
their last in 1829 and 1840. Goethe, whom Humboldt met
several times again, more than twenty years after returning
from America, was buried in 1832, as was his publisher Johann Friedrich Cotta. The death of his beloved sister-in-law
Caroline three years earlier had shaken him deeply; that of
his brother in April 1835 was even more painful. Alexander
felt closer to Wilhelm than anyone else. No one had
accompanied him for so long, so lovingly, and with so
unflinching a gaze for his strengths and weaknesses.

Although Humboldt saw his earlier companions turning into monuments around him, he remained focused on
his work. "I love what I understand, what I fully compre-
hend." With these words Humboldt ended a letter from
1839 announcing the first installment of his final work,
Cosmos.[28] A few months earlier, Humboldt had visited the
Paris studio of Louis Daguerre to inspect the metal plates
(or daguerreotypes) that would go down in history as an
early form of photography. He immediately recognized
their revolutionary significance. The aging Humboldt was
entering a new era.

7

Between Cosmos
and Fragments

THE FINAL YEARS, 1840–1859

IN PRUSSIA, the 1840s began with a new king and old challenges. Frederick William III, who had always held a protective hand over Humboldt, died in June. Humboldt even traveled to Königsberg, today's Kaliningrad, in East Prussia to attend the coronation of the king's oldest son and successor, Frederick William IV. Prussia was still a state without a constitution and basic rights for all citizens. But social problems and rising inequality in a rapidly industrializing society demanded new political solutions. The middle class, artisans, and burgeoning proletariat increasingly insisted that their claims to participate and take responsibility in political life could no longer be ignored. The hopes for liberalization that many had invested in the new king failed to materialize. Frederick William IV and the archconservative circle of advisors and military personnel with which he surrounded himself resisted far-reaching reforms. They refused to commit to a constitutional monarchy.

In the meantime, Frederick William saw mutual benefit in demonstratively keeping Humboldt, whose liberal inclinations were no secret, on side. He even tied him more closely to the royal court. In 1840, he appointed the cosmopolitan scientist to the Prussian State Council. Two years later, Frederick William named Humboldt the first chancellor of the newly established civil class of the order Pour le Mérite. In 1847, he was inducted into the Order of the Black Eagle, the highest honor Prussia could bestow. The special relationship that arose between Humboldt and the artistically and scientifically interested monarch stuck out from the rigid hierarchy of court society.

Humboldt did not gain political power from his closeness to the throne; unlike his brother, he never aspired to high office. He offered the king intellectual stimulation, and he burnished Prussia's international standing with his prestige. This became clear when Frederick William invited Humboldt to accompany him on several journeys abroad, including to the Kingdom of Denmark in 1845, and to join him on ceremonial occasions such as the festivals celebrating the ongoing construction of Cologne Cathedral in 1842 and 1848. Humboldt further exploited his direct access to the king to request financial assistance, honors, and positions for colleagues. He often succeeded in persuading Frederick William to patronize individual scientists and artists, among them such diverse figures as the

FIGURE 9. Humboldt in his library in the Oranienburger Straße in Berlin (1845). After Eduard Hildebrandt, *Humboldt in His Library*, 1856.

writer Bettina von Arnim, the Swiss naturalist Louis Agassiz, and the explorer Robert Schomburgk. Whether at the royal court or in his residence at Oranienburger Straße 67, where he had lived since 1842, Humboldt remained caught up in a round of ceaseless activity and writing.

For the conservative society at court, Humboldt's celebrity was no reason to treat him with less suspicion. One of the political issues on which Humboldt took a critical stance was Frederick William's attempt to impose renewed restrictions on the legal status of Jews in Prussia. Such discrimination went against Humboldt's humanitarian convictions and his various connections to the Jewish middle class in Berlin, dating all the way back to his youth. Upon resuming residence in Berlin, Humboldt intensified his contact with the Jewish Beer and Mendelssohn families. The bankers Joseph Mendelssohn, son of the philosopher Moses Mendelssohn, and his son Alexander took care of Humboldt's finances. By 1846, Humboldt had run out of money and repeatedly needed credit, which the Mendelssohns were willing to advance him interest-free. Humboldt also admired the composer Felix Mendelssohn Bartholdy, and he supported the opera composer Giacomo Meyerbeer. Humboldt even arranged for the unbaptized Meyerbeer to be admitted into the Prussian order Pour le Mérite. His relationship with Meyerbeer was particularly close, perhaps because the composer, like Humboldt, crossed back and forth between Berlin and Paris.

One of the journeys Humboldt made in the king's entourage brought him to London in January 1842. There he met a colleague who would soon become world-famous himself: Charles Darwin. Darwin was forty years younger than Humboldt and had already completed his five-year circumnavigation of the globe on the *Beagle* (1831–36). Darwin admired Humboldt. He had read the latter's account of the American voyage before setting out, just as Humboldt had been accompanied in South America by recollections of Georg Forster's travels. Five years prior to meeting Humboldt in person, Darwin had begun his notebook on the transmutation of species; six months after their encounter, he would finish a private sketch of his evolutionary theory. It posited the shared ancestry of all life forms (leaving open the tricky issue of human ancestry) and introduced the idea of natural selection among related species. In these years, Darwin worked on publishing his observations on coral reefs and the geology of South America.[1] He was all the more eager to hear Humboldt's thoughts about species distribution and migration.

The meeting proved a disappointment for Darwin. As so often, Humboldt talked incessantly. Still, their mutual regard endured, even if Humboldt never found his own way to the theory of evolution. While he was interested in fossil finds and geological history, he did not tackle directly the issue of evolution and preferred to concentrate, especially in his late work, on unfolding a panoramic view of the natural world. An emphasis on descriptiveness, sometimes encyclopedic in scope, characterized German natural science and geography before Darwin and was evident in Carl Ritter, too. This emphasis obscured Darwin's questions concerning the origin of species, their gradual transformation in the reproductive process, and their extension in deep time.

There were certainly parallels between Humboldt and Darwin. Both were variants of the increasingly endangered species of the 'gentleman scientist,' scholars of independent means who carried out their research outside the university system and without an institutional position. Both spent years working up the results of research trips while orchestrating a vast network of correspondence. Humboldt remained more mobile than Darwin. Unlike the English naturalist, he never married or had children. He also ranged across a wider disciplinary terrain. Humboldt was less plagued by health problems than Darwin, whereas the latter proved more adept at managing his finances. Humboldt cultivated closer ties with royal and state authorities, happily accepted their financial support, and played a personal role in influencing science policy. Since his youthful period of employment in the Prussian mining service, Humboldt was also more comfortable with research being stimulated by economic considerations.

From the 1840s, however, the position of Humboldt's science shifted in relation to the ideals and practices of the expanding research landscape that surrounded it. In Prussia and other German states, the trend for university research to become ever more specialized was accommodated through the creation of new institutional spaces. Technical institutes were established, and state-of-the-art research laboratories began to proliferate within the university. Justus von Liebig and Heinrich Gustav Magnus set up pioneering labs in Giessen and Berlin for analytical chemists and physicists, respectively. Polytechnics and trade schools flourished, along with teacher training colleges dedicated to mathematics and the natural sciences. Scientific research and education thus became ever more specialized, professional, and anchored in disciplines that, for their part, branched out into subdisciplines. Increasingly, they served the needs of an emerging industrial society. Although Humboldt, a professional in his own right, welcomed this development, his age alone prevented him from becoming a driving force in it.

The shift to analytical, laboratory-based research was particularly evident in physics, chemistry, physiology, and biology (a term that became entrenched only later). Humboldt's stance was ambivalent. He knew and supported the professionally ambitious scientists who spearheaded this development, including Liebig and the even younger Emil du Bois-Reymond, Hermann von Helmholtz, and Rudolf Virchow. He was always keen to learn from the succeeding generation. Their empiricism was congenial to Humboldt, who had never been committed to philosophical idealism. Yet Humboldt differed in his intellectual profile from the new generation of scientists as well as from the neohumanists. He persisted in wanting to demonstrate that sober scientific analysis could give rise to a holistic understanding of the world. In his view, the kind of empirical, quantitative science he had practiced all his life was compatible with the subjective and aesthetic experience of nature, to which he always remained attuned; they coexisted harmoniously in the cosmos.

Humboldt had been developing these ideas since the late 1790s. He had expanded them through his travels, tirelessly refined them in his research, and presented them to the public in his Cosmos lectures in 1827–28. Now, "late in the evening of a very active life," as Humboldt wrote,[2] it was time to bring in the harvest. The first volume of *Cosmos* appeared in 1845, the second in 1847, followed by another two volumes in the 1850s and one more posthumously. Humboldt reiterated his leitmotif, the phenomena of the earth "in their general interconnection, nature as a whole animated and moved by inner forces."[3] The main part of the first volume consisted of an "outline of a general painting of nature"[4] in which the author, as he had done in the Cosmos lectures, surveyed astronomy and geology before turning to the broad sweep of human history. In the second volume, Humboldt argued that empirical research and the aesthetic experience of nature went hand in hand. To that end, he stressed the importance of artistic strategies, especially landscape painting. In subsequent volumes, Humboldt embellished aspects of his nature painting with specialized scientific knowledge.

Like so many of Humboldt's publications, *Cosmos* was destined to remain incomplete. An accompanying atlas, illustrated by cartographer Heinrich Berghaus, appeared separately due to a falling out with Humboldt's publisher. *Cosmos* thus became a monumental torso of a "physical description of the universe."[5] Still, in hindsight, the book ranks as one of the richest products in the history of science. Once again, the author explicitly pitched his work at a general audience. The "breath of life" could not be missing, and "encyclopedia superficiality" was to be avoided.[6]

Humboldt was admirably frank in acknowledging the difficulties his project entailed. As it turned out, the details threatened to swamp the overall design. An unwieldy apparatus of footnotes and other scholarly features, along with its confusing structure, made *Cosmos* a work of bewildering complexity. The result resembled more a synopsis of a mass of materials than a coherently argued synthesis of individual findings. That achievement was reserved for Darwin, whose seminal book, *On the Origin of Species*, would appear in 1859, the year Humboldt died—a double rite of passage that marked both a passing of torches to the next generation and a paradigm shift toward the modern theory of evolution.

None of this detracted from the commercial success of Humboldt's *Cosmos*. Within four years, some 20,000 copies of the first volume were sold. For the second volume, the Cotta publishing house began with an initial print run of 10,000.[7] These were and still are spectacular figures for a science book. Humboldt had finally written a bestseller, and *Cosmos* remained in the consciousness of the reading public for generations to come. The unusually high demand led to numerous translations and reprints. Yet some contemporaries candidly pointed out that the book was difficult to consume. Well-meaning commentators acknowledged that even enthusiastic readers "dejectedly put down" *Cosmos* because it assumed too much prior knowledge.[8] The gap between Humboldt's ambition and his readers' needs was soon filled by self-appointed popularizers. They began publishing simplified versions and explanations of the work. A veritable "Cosmos literature"[9] emerged in Humboldt's wake.

Humboldt's deliberate attempt to break down the barriers between disciplinary science and the public had taken on a momentum that the author could no longer control. For all his justified pride in his success, the dilemma was obvious. The title *Cosmos* was unprotected, and Humboldt had no wish to see his life's work jeopardized. More importantly, readers demanded more accessible formats. In this situation, Humboldt tried to make a virtue of necessity. From 1848, he worked with Cotta to bring out a briefer, streamlined, and elegantly written *Micro-Cosmos*. This would have the agreeable side effect of paying off financially and helping him get back in the black.[10]

Although the project fell through, Humboldt deserves credit for having recognized the need for new forms of popular science to disseminate scientific education. He acknowledged that some among the new breed of popularizers had the competence required for the task. Foremost among them was the versatile zoologist Emil Adolf Rossmässler, a pioneer of popular science in Germany. Through traveling lectures, new periodicals, and easily digestible books on the natural history of the universe, Rossmässler, his friend Otto Ule, and others helped bring Humboldt's work to a wider audience. They advocated a participatory understanding of research and addressed readers who had been relegated to the status of amateurs through the increasing professionalization of the scientific community. At Rossmässler's initiative, Humboldt societies and Humboldt festivals were founded in 1859 to promote scientific education for the public and link it to a democratic idea of national education.[11]

It was no accident that Humboldt's admirers Rossmässler and Ule joined the 1848 revolution on the side of the democrats and in support of a constitutional state. In the second half of the 1840s, when critics of political oppression in Germany rebelled and created new spaces for communication, modern scientific thinking provided a vehicle for questioning the established social order. Many popularizers embraced the natural sciences as a countermodel to tradition and state authority. In their commitment to rational analysis, empirical evidence, and logical conclusions, which were ultimately meant to generate a 'natural *Welt-anschauung*,' they invoked Humboldt as Germany's most celebrated naturalist. It helped that Humboldt had no religious affiliations. He was known for his liberal inclinations and had acquired a reputation early on as a staunch critic of colonial subjugation.

As so often, however, Humboldt avoided taking a clear political stand when the revolution broke out. In 1848, just a few weeks after he had returned from his final stay in Paris, the February Revolution in France forced the abdication of King Louis Philippe. In the German states and Vienna, too, demands for political change and constitutions grew louder. On March 18, an initially peaceful gathering in front of the royal palace led to widespread unrest on the streets of Berlin. The military stormed the barricades erected by a disaffected coalition of middle-class Berliners, workers, and artisans. Hundreds died or were injured in the fighting. Frederick William IV now changed course. Seeking to deescalate the conflict, he made a symbolic gesture. Along with other members of the Royal Academy and the aristocracy, Humboldt joined the funeral procession in which the coffins of the 'March Fallen' were borne past the king to the burial site. This was a stage show put on to pacify the revolutionaries, not a declaration of solidarity. On the same day, Humboldt noted in a letter to Gabriele von Bülow, his late brother's daughter, how impressed he had been by the calm mood of the crowd. He had heard "no hint of indecency, no word of politics" and been treated respectfully by the public.[12]

As events unfolded, Humboldt watched with a mixture of sympathy and skepticism. From May, the Prussian National Assembly convened for several months in the Singakademie, where he had lectured on the cosmos two decades before. Meanwhile, the German National Assembly sat in Frankfurt am Main. Both debated principles that were dear to Humboldt, such as freedom of the press, freedom of religion, and constitutional government. Besides, he had little time for the Christian Romanticism of his king. Like many liberals and conservatives, he never overcame his fear of a social revolution and what he saw as the "dark"[13] undertone of uncontrollable mass movements. An enlightened, reform-oriented, and constitutional monarchy tied to a bourgeois legal order presumably remained his ideal for both Prussia and Germany.

Humboldt looked on as the Prussian government declared a state of emergency in Berlin in late fall, driving the National Assembly from the city. The king imposed a constitution on the country. In April 1849, Frederick William IV turned down the German National Assembly's offer for him to be crowned emperor in a new, constitutionally governed German Empire. Keen to demonstrate his personal loyalty to the king, Humboldt voiced his approval of the decision. The exclusion of Austria, and the possible affront to the sizable Catholic population, stood opposed to the idea of a loose pan-German confederation that Humboldt seems to have favored.

When Prussia took a reactionary turn after 1849, Humboldt stayed true to his convictions. In 1853, he supported a petition against restrictions on religious freedom in

Prussia. Two years later, he publicly protested that an English translation of his work on Cuba had omitted passages critical of slavery. It was only fitting that in 1857, Humboldt supported the passage of a law granting freedom to any enslaved person who set foot on Prussian soil. The rise of the industrial world, closely tied to the achievements of the modern sciences, continued to fascinate him. In Berlin, he followed the success of the Siemens & Halske telegraph company as well as the Borsig engineering works, which produced their one thousandth locomotive in 1858.[14] A new age was dawning.

In these tumultuous times, Humboldt celebrated his eightieth birthday in 1849 in the presence of the royal couple. He found himself increasingly forced to slow down his pace. His remaining energies were devoted to the unfinished volumes of *Cosmos*, the ultimately unrealized plan for *Micro-Cosmos*, and the third edition of *Views of Nature*, published in 1849. The preparation of his *Kleinere Schriften* (*Shorter Writings*, 1853) and his efforts to assemble "Fragments on the Geography of Plants"[15] also took their toll. Humboldt was visibly concerned for his posthumous reputation, even as he joked with his publisher that he was "only half-dead."[16] He sensed that his attempt at a cosmic synthesis was beginning to overwhelm both him and his readers. Not by chance, his remarks in *Cosmos* that he could only present his complex themes in a "fragmentary" manner became more frequent.[17]

For all that, Humboldt continued to pursue topics that had interested him for decades, such as geomagnetism and the construction of a canal leading through Panama, connecting the Pacific Ocean and the Caribbean Sea. He still sought to do justice to whatever struck him as new and important. For example, he incorporated observations by Austrian globetrotter Ida Pfeiffer and Scottish mathematician Mary Somerville into the *Cosmos* volumes.[18] He continued to support emerging talents and receive visitors from around the world, many of them from the United States. He was intrigued by the African continent and by the expeditions undertaken there by younger researchers. It is also striking how Humboldt came to relish the company of artists in his twilight years. Along with the sculptor Christian Daniel Rauch and the composer Franz Liszt, he befriended several landscape painters, in keeping with his long-standing fondness for this branch of the visual arts. At the royal court and elsewhere, Humboldt had become a living legend. Among family members, he referred to himself only half in jest as "Uncle Cosmos."[19] He had already bequeathed his personal estate years earlier to his servant, Johann Seifert.

Meanwhile, the flood of requests for advice, personal meetings, and mementos from the wizened scholar continued unabated. Humboldt could no longer keep up. In February 1857, he likely suffered a mild stroke. More than any other scientist before him, Humboldt had become the center of a far-flung, transcontinental network of contacts. He had shouldered an immense correspondence and spent much of his life exposed to the glare of public attention. In March 1859, he felt obliged to publish a "plea for help" in the *Berlinische Zeitung*. He politely asked everyone to allow him rest in light of his "waning physical and mental powers."[20]

On May 6, 1859, Humboldt passed away in his Berlin residence on Oranienburger Straße, four months short of his ninetieth birthday. After the public viewing of his body at his home, a solemn funeral procession took place four days later, followed by a state ceremony at Berlin Cathedral. From there the coffin was conveyed to Schloss Tegel, where Alexander von Humboldt was laid to rest on May 11, 1859, in the park of the castle where he had spent much of his childhood and youth.

———

Humboldt has since been repeatedly summoned from the grave. From 1859 to now, interpretations of his personality and work have catered to ever-changing contemporary concerns, forming a "metabiography" that reflects both Humboldt's versatility and the attempts made by subsequent generations to claim his legacy.[21] In 1859, Rossmässler was trying to elevate the recently interred scientist into a national figure alongside Friedrich Schiller, whose hundredth birthday was celebrated in the same year.[22] The following year saw the establishment of the Alexander von Humboldt Foundation for Natural Science and Travel. The Alexander von Humboldt Foundation in its present-day form, relaunched in West Germany in 1953, continues Humboldt's legacy globally by supporting scientists and scholars to come to Germany.

Humboldt was too much a global citizen to be made the posthumous property of any nation-state. The attempt to honor him with a national monument in Germany demonstrated this ambivalence. It ended in 1883 with the unveiling of two Humboldt memorials in front of Berlin University, one for Alexander, the other for his brother Wilhelm. Today, Humboldt can be found commemorated in busts and monuments from Tenerife, the first stop on his American journey, to Mexico and New York and back to Berlin. German immigrant communities in the United States were behind the particularly imposing statues erected in Philadelphia, St. Louis, and Chicago.[23] Travelers to Hispanic America, where Humboldt has been venerated for two centuries, will stumble across him everywhere. Around the world, Humboldt has lent his name to rivers and counties, mountain ranges and towns, and even a mare on the moon, reflecting the cosmic reach of his intellect.

In light of renewed trends to turn Humboldt into a timeless hero serving our cultural preferences or, conversely, to ascribe to him a colonial mindset in response to modern critique, it is worth appreciating the various and conflicting aspects of his legacy. Some are tied to the historical circumstances that framed his thinking, others reach into our present and retain a remarkable freshness. Humboldt himself came to feel toward the end of his life that he had not fully realized his potential. He regarded his work on plant geography, isothermal lines, and the earth's magnetic field as his most important contributions to the history of science.[24] Many more insights and achievements could be added to

FIGURE 10. Statue of Alexander von Humboldt in St. Louis, Missouri (1878).
© Andreas W. Daum, Private Archives.

this list. For example, Humboldt triggered the international use of guano as a fertilizer beyond long-standing practices in South America, he measured stellar luminosity in the Southern Hemisphere with greater precision than most of his predecessors, and he did valuable work measuring altitudes. He left behind intriguing observations on the currency systems and ore deposits of his time, delved into speleology and botany, and made instructive drawings of flora and fauna. The "Humboldtian gaze" as "a way of seeing that was at once morphological and numerical, aesthetic and scientific, local and global" is one that still captivates.[25]

The tension Humboldt experienced between his quest for a grand synthesis and his forced retreats into the fragmentary is evident in the sprawling and confusing character of his published oeuvre. Yet this dilemma was far from an individual problem. It encapsulated the epistemological challenge faced by an age that saw a true explosion of knowledge—a challenge we live with even more intensively in the twenty-first century. Unlike Darwin, Humboldt did not come up with a clearly defined theory that fundamentally changed scientific and social thinking. Humboldt was neither a revolutionary in the scholarly realm nor one in the political world. But he left us with myriad complex thoughts and incentives for further research.

Finding contradictions in Humboldt's oeuvre and his biography can only surprise us if we subscribe to an illusory longing for biographical coherence or an ahistorical presentism. What invites criticism today from a postcolonial perspective should be as much a part of any fair assessment of Humboldt as what makes him progressive: his humanitarian interest in people and his critique of colonial suppression; his ecological awareness and his interest in connecting empirical observation and aesthetic perception; his efforts to raise public awareness of science and his determination to acquire knowledge on a global scale; and—last but not least—his efforts to compare and disseminate scientific observations in transcultural networks.

Precisely the transdisciplinary reach and almost unruly, open-ended character of Humboldt's science have allowed it to exert a uniquely broad influence, securing him a lasting place in the cultural history of modernity. To be sure, Humboldt's scientific studies, measurements, and country profiles, complemented by his historical, social, and ethnological observations, do not always conform to today's standards of scholarship. They contain errors and have been surpassed by succeeding generations of researchers (no surprise here either). For all that, they are extraordinarily rich and have had a varied impact on different continents. Humboldt was instrumental in transforming the natural sciences into a powerful paradigm for understanding the world and our place in it. At the same time, he transcended the supposed opposition between the "two cultures," the natural sciences and the humanities.[26] He became a mediator between science, literature, and the arts. His idea of a cosmos in which everything—nature, culture, society—is interconnected has lost none of its appeal, not least due to its very incompleteness. Besides popular science, it has left its mark on landscape painting, literature, and environmentalism in both Europe and America.

The Humboldt effect lingers. No matter what topic we are interested in, there is always something to find in Humboldt. Above all, he invites us to be perpetually curious and to create a society that fosters communication among different cultures. Humboldt broadened the horizons of his society. He guided his readers and listeners on a tour through Europe and the Americas to Asia and even Africa, despite never having visited this continent himself. In his thinking and through his travels, he linked dispersed intellectual and geographical locations, forging new connections around the world. Humboldt made clear that this process—today we know it as globalization and intercultural exchange—is never-ending. In drawing diverse physical spaces, cultures, and continents into the embrace of his boundless curiosity, Humboldt emerges as a historical figure who was deeply embedded in his times yet still speaks to us today.

CHRONOLOGY

Geographical terms have been modernized and adapted to today's borders. Book titles are indicated in italics.

1769	September 14: birth of Alexander von Humboldt
1779	Death of father, Alexander Georg von Humboldt
1787	Until 1788, student at the University of Frankfurt/Oder
1788	Return to Tegel and Berlin
1789	Until 1790, student in Göttingen
	Study trip to the Rhine region with S. J. van Geuns
1790	*Mineralogical Observations on Some Basalts at the Rhine*
	Journey with Georg Forster to the Rhineland,
	Belgium, the Netherlands, England, and France
	Until 1791, student at the Trade Academy in Hamburg
1791	Until 1792, student at the Mining Academy in
	Freiberg, Saxony
1792	From March in the Prussian mining service
	Travels to Bavaria, Austria, and Poland
1793	Until late 1796, chief inspector of mines in Franconia
	Florae Fribergensis specimen
1794	*Aphorisms from the Chemical Physiology of Plants*
	Travels to Pomerania, Posen, and Bohemia, and on a
	diplomatic mission to the Rhineland
	Until 1797, periodic visits to Jena
1795	Essay on the "Rhodian Genius" in Schiller's *Horen*
	Journey to the Alps in Austria and Switzerland
1796	Death of mother, Maria Elisabeth

1825	Trip to the Bretagne
1826	*Political Essay on the Island of Cuba*
	Second edition of *Views of Nature*
1827	Return to Berlin
	Winter to spring 1828: Cosmos lectures at Berlin
	University and in the Berlin Singakademie
1828	Cohost of the seventh conference of the Society of
	German Natural Scientists and Physicians in Berlin
1829	Journey to the Russian Empire and Central Asia
1830–48	Eight stays in Paris
1831	Final meeting with Goethe in Weimar
	Fragments on the Geology and Climatology of Asia
1834	Until 1839, *Critical Investigations into the Historical*
	Development of Geographical Knowledge Respecting the
	New World
1835	April 8, death of Wilhelm von Humboldt
	Until 1847, diplomatic reports from Paris
1838	Visit to Daguerre's studio in Paris
1840	To Königsberg (Kaliningrad) for the coronation of
	Frederick William IV
1842	Journey to England, meeting with Charles Darwin
	Chancellor of the civil class of the Prussian order
	Pour le Mérite
1843	*Asie centrale*
1845	First volume of *Cosmos*, the following volumes,
	1847 to 1862
	Visit to Denmark
1847	Knight of the Prussian Order of the Black Eagle
	Final journey to Paris
1848	Funeral procession for the fallen of the revolution
	in Berlin
1849	Third edition of *Views of Nature*
1853	*Shorter Writings*
1858	Fourth volume of *Cosmos*
1859	May 6, death in Berlin
	Burial in the gardens at Schloss Tegel

GUIDE TO SOURCES
AND FURTHER READING

A large number of sources document Alexander von Humboldt's life and works. They are, however, scattered all over the world and not easy to survey. Humboldt wrote and received tens of thousands of letters—all handwritten and often difficult to decipher; no typewriters or computers yet. Moreover, research pertaining to Humboldt raises a considerable linguistic challenge. Much of this correspondence is written in German and French, as are his publications. Even though this massive pool of sources is critical to understanding Humboldt, it requires translations to become globally accessible and to fully appreciate a man who connected different cultures and conversed in several languages.

Chunks of Humboldt's vast correspondence have been published, beginning with several hundred of his *Jugendbriefe*, "Letters from His Youth," covering the years 1787 to 1799. These are available in an eponymous edition by Ilse Jahn and Fritz G. Lange (1973). Meanwhile, almost fifty volumes of Humboldt's communications with contemporaries—fellow scientists, his publisher Cotta, the Prussian king, and others—have been included in

the series "Beiträge zur Alexander-von-Humboldt-Forschung"; see the (incomplete) list at https://www
.degruyter.com/serial/bahf-b/html. Since 2015, a project
associated with the Berlin Academy of Sciences has
worked on publishing, both in print and digitally, Humboldt's travel accounts, including diaries, drawings, and
correspondence; see https://www.bbaw.de/forschung
/alexander-von-humboldt-auf-reisen-wissenschaft-aus
-der-bewegung. The digital journal *HiN: Alexander von
Humboldt im Netz* informs readers about this project and
new research. For previous and still valuable editions of
Humboldt's American diaries, see the volumes, all edited
by Margot Faak, Alexander von Humboldt, *Reise durch
Venezuela* (2000); Alexander von Humboldt, *Reise auf
dem Río Magdalena, durch die Anden und Mexico* (1990,
second edition 2003); and *Alexander von Humboldt: Lateinamerika am Vorabend der Unabhängigkeitsrevolution. Eine
Anthologie von Impressionen und Urteilen aus seinen Reisetagebüchern* (2003).

The corpus of texts Humboldt himself published is no
less confusing. He put out a great deal, starting with small,
scholarly notes and ending in voluminous treatises. At
times, his publications consisted of individual parts released over several years. Many were later revised by him
in new editions, and some remained incomplete. A few
included both a German and French version, such as his
Geography of Plants. Several of Humboldt's publications
involved different contributors, which is not always clear
from the titles. Moreover, Humboldt himself and all
readers since have struggled with competing translations

of his works. The only reliable bibliography, edited by Horst Fiedler and Ulrike Leitner (*Alexander von Humboldts Schriften*, 2000), includes translations and is an indispensable tool for all Humboldt research. Here, too, patience is needed; this volume encompasses almost five hundred pages.

Hanno Beck, who wrote a still important Humboldt biography in German (1959–61), has compiled several of Humboldt's texts in a reader-friendly, seven-volume compendium, *Alexander von Humboldt: Darmstädter Ausgabe* (second edition 2008). In ten volumes, the *Berner Ausgabe* (2019) makes available those of Humboldt's publications that appeared outside the book format, such as journal articles. Other recent publications include Humboldt's drawings, more travel accounts, and more correspondences. Most, if not all, of Humboldt's book publications have been digitized and are today accessible via electronic catalogues.

Many of these publications have not been translated into English. Most older translations, for example, of *Cosmos* and Humboldt's narrative of his American journey, are either abbreviated, unreliable, or both. As Alison Martin (2018) has shown, translating Humboldt into English has always been a process of restyling the author's persona and authority, adjusting his work to changing cultural preferences. For *Essay on the Geography of Plants*, see the new edition by Stephen T. Jackson (2009), for *Views of Nature*, that by Jackson and Laura D. Walls (2014), and for Humboldt's works on Cuba, the Cordilleras, and New Spain, the series Alexander von

Humboldt in English, edited by Vera Kutzinski and Ott-mar Ette. Humboldt's account of his American journey until 1801 is abridged in *Personal Narrative* (1995) and available in long versions in the older translations by Helen Maria Williams (1814–29) and Thomasina Ross (1852–53).

It is encouraging that authors from various disciplines have taken interest in Humboldt's oeuvre in the last three decades. Focusing on English-language publications, this short survey is meant to help readers in exploring Humboldt and the era in which he lived. The growing output of articles and chapters dealing with Humboldt's science can be followed best through the *Isis Current Bibliography*, published annually and accessible electronically. Nicolaas A. Rupke analyzes previous interpretations and appropriations of Humboldt in his overview, *Alexander von Humboldt: A Metabiography* (2008). A comprehensive biography is still missing. Older, short English-language portrayals of Humboldt's life, such as those by Helmut de Terra (1955), L. Kellner (1963), Douglas Botting (1973), and Donald McCrory (2010), are imbalanced in their preference for Humboldt's American journey and do not reflect up-to-date scholarship. Andrea Wulf's *The Invention of Nature* (2015) and Maren Meinhardt's *Alexander von Humboldt* (2019) offer popular, heroic accounts. Unsurpassed among the biographies of Humboldt's contemporaries are Janet Browne's *Charles Darwin* (2 vols., 1994–2003) and Nicholas Boyle's *Goethe* (1991–2000). For Georg Forster, see the studies by Jürgen Goldstein (2019) and Todd Kontje (2022).

Easy access to major themes in European history during Humboldt's times is provided by *Companion to Eighteenth-Century Europe*, edited by Peter H. Wilson (2014), and *Nineteenth Century Europe*, edited by Stefan Berger (2009). For Prussia and Germany, see Christopher Clark, *Iron Kingdom* (2006), David Blackbourn, *History of Germany, 1780–1918* (2003), and Helmut Walser Smith, *Germany: A Nation in Its Time* (2020), as well as *The Oxford Handbook of Modern German History* (2011), edited by H. W. Smith. The Holy Roman Empire, the Napoleonic Wars, and the ensuing transformation of Germany are explained in Peter H. Wilson's *Heart of Europe* (2016), Joachim Whaley's *Germany and the Holy Roman Empire* (vol. 2, 2012), Brian E. Vick's *The Congress of Vienna: Power and Politics after Napoleon* (2014), and Brendan Simms's *The Struggle for Mastery in Germany* (1998). The transformation of Berlin in Humboldt's lifetime can be traced through Alexandra Richie's *Faust's Metropolis* (1998) and the two volumes of Wolfgang Ribbe's *History of Berlin* (in German, third edition 2002).

Readers interested in the global history of Humboldt's epoch, including that of colonial empires and slavery, may turn to Jürgen Osterhammel's magnificent *Transformation of the World* (2014), volumes 3 and 4 of *A History of the World*, edited by Akira Iriye and Osterhammel (2015–18), and the *Companion to Latin American History*, edited by Thomas H. Holloway (2011). David Blackbourn's *Germany in the World* (2023) offers a superb history of Germans' interactions with European and non-European worlds, with due emphasis on traveling naturalists. The essays in *The Brokered World: Go-Betweens and Global Intelligence, 1770–1820* (2009), edited by Simon Schaeffer et al., highlight the role of intermediators in knowledge transfers. Aaron Sachs (*Humboldt Current*, 2006), Laura D. Walls (*The Passage to Cosmos*, 2009), Sandra Rebok (*Humboldt and Jefferson*, 2014), and Eleanor J. Harvey (ed., *Alexander von Humboldt and the United States*, 2020) address the repercussions of Humboldtian ideas in the United States.

Research on science's changing concepts, practices, and agents from the early modern era has flourished in the last decades. *The Companion to the History of Science*, edited by Bernard Lightman (2020), allows convenient access to a wide range of topics, as do volumes 3 to 8 of the *Cambridge History of Science*, edited since 2008 by David C. Lindberg and Ronald L. Numbers. Seminal studies by Lorraine Daston and Londa Schiebinger open windows into early modern and modern knowledge making. Kathryn Olesko (1991), Arleen Tuchman (1993), Lynn K. Nyhart (1995 and 2009), and William Clark (2006) portray the role of science and universities in Germany from different perspectives in their respective monographs. Robert J. Richards (2004 and 2016), Peter Reill (2005), and Dorinda Outram (2019) expand on the changing meanings of science and nature in essential books on the Enlightenment and the Romantic era. The essays in *The Sciences in Enlightened Europe*, edited by William Clark, Jan Golinski, and Simon Schaffer (1999), shed light on the complexity of this epoch.

The history of scientific traveling and naturalists' role in colonial explorations is the subject of many works; see only Annick Foucrier, ed., *The French and the Pacific World* (2003), Harry Liebersohn, *The Traveler's World* (2006), and David Igler, *The Great Ocean* (2013). Colonial and transnational aspects—within and beyond the European empires—are emphasized in *Visions of Empire*, edited by David P. Miller and Peter H. Reill (1996), and by Susanne Zantop, *Colonial Fantasies* (1997), Russell Berman, *Enlightenment or Empire* (1998), Richard Drayton, *Nature's Government* (2000), Jorge Cañizares-Esguerra, *Nature, Empire, and Nation* (2006), Neil Safier, *Measuring the New World* (2008), Mary Louise Pratt, *Imperial Eyes* (2009), Daniela Bleichmar et al., eds., *Science in the Spanish and Portuguese Empires* (2009), John R. McNeill, *Mosquito Empires* (2010), and in *Global Scientific Practice in an Age of Revolutions*, edited by Patrick Manning and Daniel Rood (2016). Aaron Sachs's article on the postcolonial take on Humboldt in *History and Theory* 42 (2003) provides a thoughtful corrective to an overdrawn critique. Londa Schiebinger's "Forum Introduction: The European Colonial Science Complex" contains a summary of recent trends in scholarship. For discussion about the history of Iberian, Hispanic, and Creole science, see Gregory Cushmann, "Humboldtian Science," *Osiris* (2011) and the essays in *The Invention of Humboldt*, edited by Mark Thuner and Jorge Cañizares-Esguerra (2023). Michael Zeuske has dealt extensively with Humboldt in the Caribbean world and his take on slavery.

Regarding the scientific disciplines that were taking shape during Humboldt's life, readers will profit from consulting the books by Martin J. S. Rudwick on the history of geology and those by Peter J. Bowler on concepts of evolution and Darwinism. Still valuable is Ernst Mayr's *The Growth of Biological Thought* (reprint, 2003). On botany, see Daniela Bleichmar, *Visible Empire* (2012) and H. Walter Lack, *Alexander von Humboldt* (2018). The essays in *Alexander von Humboldt: Multiperspective Approaches*, edited by Gregor Falk, Manfred Strecker, and Simon Schneider (2022), illuminate various aspects of Humboldt's science. In recent years, historians have paid increasing attention to science's role in the public sphere, the emergence of popular science, and variations of amateur science. My book *Wissenschaftspopularisierung im 19. Jahrhundert* (Popularizing Science in the Nineteenth Century) provides a comprehensive account of developments in Germany (1998; second edition 2002), while my essay "Varieties of Popular Science," *Isis* 100 (2009): 319–32, informs readers about the emergence of popular science in other countries. A panorama of popular activities is displayed in *Cultures of Natural History*, edited by N. Jardine, J. Secord, and E. C. Sparry (1996).

The critical role of technology, artisanal industrial production, and growing state and consumer demands is explained by Andre Wakefield, *Disordered Police State: German Cameralism as Science and Practice* (2009), Ursula Klein, *Humboldts Preußen* (in German, 2015), and Suzanne L. Marchand, *Porcelain* (2020). James J. Sheehan's *German History, 1770–1866* (1989) and Thomas Nipperdey's

Germany from Napoleon to Bismarck (1996), augmented by comparative and global perspectives in the volume *The Global Bourgeoisie*, edited by Christof Dejung, David Motadel, and Jürgen Osterhammel (2019), illuminate the middle-class culture that began to embrace the natural sciences and provided fertile ground for appreciating Alexander von Humboldt.

ACKNOWLEDGMENTS

This book has emerged from my ongoing work on Alexander von Humboldt and his biography. I thank all those institutions that have supported my research along the way, including the Humanities Institute at the State University of New York (SUNY) at Buffalo with a Faculty Fellowship in 2013, the National Endowment for the Humanities with a Research Fellowship in 2014–15, and the Smithsonian Institution in Washington, D.C., where the Baird Resident Scholarship in 2017 allowed me to dig deep into Humboldt's original publications.

My special thanks go to the Alexander von Humboldt Foundation for the Humboldt Research Prize in 2019–20, which made an extended stay in Germany possible and facilitated numerous exchanges with fellow Humboldtians from various disciplines. During that year, I profited greatly from discussions following presentations at Cambridge University's Department of History and Philosophy of Science, the Berlin-Brandenburg Academy of Sciences, and the Societat Catalana de la Història de la Ciència i de la Tècnica in Barcelona, as well as the Deutsches Museum and the Siemens Foundation in Munich, the Ludwig Maximilians University Munich, and the University of Hamburg.

I warmly thank Dr. Ingo Schwarz (Berlin) and Professor Reinhard Stauber (Klagenfurt) for reviewing the original manuscript and their valuable feedback, as well as Eric Crahan at Princeton University Press for his early interest in this book and Robert Savage for his translation. Professor Susanne Renner (St. Louis) and Professor Andreas Kappeler (Vienna) kindly provided additional advice.

My most profound thanks go to Evis, my wife, and our children, Nicholas and Alexander, for their patience and much more.

NOTES

Introduction

1. On appropriations and interpretations of Humboldt over time, catering to ever-changing cultural preferences, see Nicolaas A. Rupke, *Alexander von Humboldt: A Metabiography* (Chicago: University of Chicago Press, 2008). Cf. Mary Louise Pratt, *Imperial Eyes: Travel Writing and Transculturation*, 2nd ed. (London: Routledge, 2009), and Andrea Wulf, *The Invention of Nature: Alexander von Humboldt's New World* (New York: Knopf, 2015).

2. Susan Faye Cannon, "Humboldtian Science," in *Science in Culture: The Early Victorian Period* (New York: Science History Publications, 1978), 73–110; Andreas W. Daum, "Humboldtian Science and Humboldt's Science," *History of Science* 62 (2024).

3. Jürgen Osterhammel, *The Transformation of the World: A Global History of the Nineteenth Century*, trans. P. Camiller (Princeton: Princeton University Press, 2014).

Chapter 1: Training the Mind

1. Thomas Nipperdey, *Germany from Napoleon to Bismarck: 1800–1866*, trans. Daniel Nolan (Princeton: Princeton University Press, 1996), 89–90.

2. Alexander von Humboldt to Carl Freiesleben, October 10, 1796, in *Die Jugendbriefe Alexander von Humboldts: 1787–1799*, ed. Ilse Jahn and Fritz G. Lange (Berlin: Akademie, 1973), 528.

3. A. von Humboldt to Carl Freiesleben, c. November 25, 1796, in *Jugendbriefe*, 553.

4. Alexander von Humboldt, *Aus meinem Leben: Autobiographische Bekenntnisse*, ed. Kurt-R. Bierman, 2nd ed. (Munich: C. H. Beck, 1989), 38.

5. A. von Humboldt to Carl Freiesleben, October 21, 1793, in *Jugendbriefe*, 280.

6. Alexander von Humboldt, *Versuche über die gereizte Muskel- und Nervenfaser nebst Vermuthungen über den chemischen Process des Lebens in der Thier- und Pflanzenwelt*, vol. 2 (Berlin: Rottmann, [1799]), 368, 398.

7. Wilhelm von Humboldt, "Bruchstück einer Selbstbiographie," in *Autobiographische Dichtungen, Briefe*, Werke in fünf Bänden, ed. Andreas Flitner and Klaus Giel, vol. 5 (Darmstadt: Wissenschaftliche Buchgesellschaft, 2010), 8–10.

8. Immanuel Kant, "Beantwortung der Frage: Was ist Aufklärung?" [1784], in *Was ist Aufklärung? Aufsätze zur Geschichte und Philosophie*, ed. Jürgen Zehbe, 3rd ed. (Göttingen: Vandenhoeck & Ruprecht, 1985), 55. Cf. Immanuel Kant, "An Answer to the Question: What Is Enlightenment?," in *Practical Philosophy*, ed. Mary J. Gregor, Cambridge Edition of the Works of Immanuel Kant (Cambridge: Cambridge University Press, 1996), 17, with a slightly different translation.

9. A. von Humboldt, *Aus meinem Leben*, 34.

10. A. von Humboldt to Wilhelm Gabriel Wegener, March 27, 1789, in *Jugendbriefe*, 47.

11. Andreas W. Daum, "Alexander von Humboldt am Rhein: Zur regionalen Grundlage von Humboldts Wissenschaft, Reisen und Politikverständnis 1789–1848," *Rheinische Vierteljahresblätter* 85 (2021): 148–84.

12. Georg Forster, *Ansichten vom Niederrhein, von Brabant, Flandern, Holland, England und Frankreich im April, Mai und Juni 1790*, ed. Gerhard Steiner (Frankfurt am Main: Insel, 1989), 156.

13. A. von Humboldt, *Aus meinem Leben*, 53.

14. A. von Humboldt, *Aus meinem Leben*, 53.

15. A. von Humboldt to Wilhelm Gabriel Wegener, September 23, 1790, in *Jugendbriefe*, 106–7.

16. A. von Humboldt to Dietrich L. G. Karsten, August 25, 1791, in *Jugendbriefe*, 144.

17. Kurt-R. Biermann, "Die Gebrüder Humboldt an der Universität Frankfurt (Oder)," in *Miscellanea Humboldtiana* (Berlin: Akademie, 1990), 48.

Chapter 2: Constantly on the Move

1. Johann Wolfgang von Goethe, "Hermann und Dorothea," in *Goethes Werke: In zwölf Bänden* (Berlin: Aufbau, 1974), 2:408. Cf. *Goethe's Hermann and Dorothea*, trans. Edgar Alfred Browning (Philadelphia: McKay, 1898), 64.

2. A. von Humboldt, *Aus meinem Leben*, 55.

3. Alexander von Humboldt, *Ueber die unterirdischen Gasarten und die Mittel, ihren Nachtheil zu vermindern: Ein Beytrag zur Physik der praktischen Bergbaukunde* (Braunschweig: Vieweg, 1799), iv.

4. A. von Humboldt, *Ueber die unterirdischen Gasarten*, 31.

5. A. von Humboldt, *Ueber die unterirdischen Gasarten*, 251.

6. A. von Humboldt to Carl Freiesleben, November 21, 1794, in *Jugendbriefe*, 379.

7. A. von Humboldt to Carl Freiesleben, November 21, 1794, in *Jugendbriefe*, 378.

8. A. von Humboldt to Archibald Maclean, February 9, 1793, in *Jugendbriefe*, 233.

9. A. von Humboldt to Karl Ludwig Willdenow, December 20, 1796, in *Jugendbriefe*, 560.

10. Andreas W. Daum, "Social Relations, Shared Practices, and Emotions: Alexander von Humboldt's Excursion into Literary Classicism and the Challenges to Science around 1800," *Journal of Modern History* 91 (March 2019): 1–37, see especially 28–31.

11. A. von Humboldt, *Versuche über die gereizte Muskel- und Nervenfaser*, 2:434.

12. Cannon, "Humboldtian Science." Cf. Michael Dettelbach, "Humboldtian Science," in *Cultures of Natural History*, ed. N. Jardine, J. Secord, and E. C. Sparry (Cambridge: Cambridge University Press, 1996), 287–304; Kathryn Olesko, "Humboldtian Science," in *The Oxford Guide to the History of Physics and Astronomy*, ed. John L. Heilbron (New York: Oxford University Press, 2005), 159–62; and Daum, "Humboldtian Science and Humboldt's Science."

13. A. von Humboldt, *Versuche über die gereizte Muskel- und Nervenfaser*, 2:436. Cf. Friedrich Schiller, "Der Spaziergang," in *Friedrich Schiller: Sämtliche Werke*, vol. 1: *Gedichte – Dramen 1*, ed. Albert Meier, 3rd ed. (Munich: Hanser, 2004), 232, and *The Poems of Schiller*, trans. Edgar Alfred Bowring (New York: Lovell, 1891), 204.

14. Peter-André Alt, *Friedrich Schiller*, 2nd ed. (Munich: C. H. Beck, 2009), 83.

15. Alexander von Humboldt, "Die Lebenskraft oder der Rhodische Genius: Eine Erzählung," *Die Horen: Eine Monatsschrift* 2, no. 5 (1795): 90–96; later incorporated in Alexander von Humboldt, *Ansichten der Natur, mit wissenschaftlichen Erläuterungen*, 2nd ed., vol. 2 (Stuttgart: Cotta, 1826), 187–200, as well as this book's third edition (1849), vol. 2, 297–314. Cf. Alexander von Humboldt, *Views of Nature*, trans. Mark. W. Person, ed. Stephen T. Jackson and Laura Dassow Walls (Chicago: University of Chicago Press, 2014), 261–64, based on the third German edition.

16. Daum, "Social Relations, Shared Practices, and Emotions."

17. Wilhelm von Humboldt, "Ueber den Geschlechtsunterschied und dessen Einfluss auf die organische Natur," [*Die Horen*, 1795] reprinted in W. von Humboldt, *Schriften zur Anthropologie und Geschichte*, Werke in fünf Bänden, ed. Andreas Flitner and Klaus Giel, vol. 1 (Darmstadt: Wissenschaftliche Buchgesellschaft, 2010), 291.

18. A. von Humboldt to Reinhard von Haeften, January 1–4, 1796, in *Jugendbriefe*, 477.

19. *Alexander von Humboldt und Cotta: Briefwechsel*, ed. Ulrike Leitner (Berlin: Akademie, 2009), 39.

20. A. von Humboldt to Friedrich von Schuckmann, May 14, 1797, in *Jugendbriefe*, 580.

21. A. von Humboldt, *Aus meinem Leben*, 56.

22. A. von Humboldt, *Aus meinem Leben*, 97.

23. Daniela Bleichmar, *Visible Empire: Botanical Expeditions and Visual Culture in the Hispanic Enlightenment* (Chicago: University of Chicago Press, 2012), 18.

24. A. von Humboldt to Karl M. E. Frh. von Moll, June 5, 1799, in *Jugendbriefe*, 682.

Chapter 3: The Interaction of All Forces

1. Wilhelm von Humboldt to Karl Gustav von Brinkmann, March 18, 1793, in *Gespräche Alexander von Humboldts*, ed. Hanno Beck (Berlin: Akademie, 1959), 6.

2. A. von Humboldt to Karl M. E. Frh. von Moll, June 5, 1799, in *Jugendbriefe*, 682.

Chapter 4: Gaining a Picture of the Whole

1. Alexander von Humboldt, *Reise in die Äquinoktial-Gegenden des Neuen Kontinents*, ed. Ottmar Ette, vol. 1 (Frankfurt a.M.: Insel, 1991), 502. Cf. Alexander de Humboldt and Aimé Bonpland, *Personal Narrative of Travels to the Equinoctial Regions of the New Continent, during the Years 1799–1804*, trans. Helen Maria Williams, vol. 3 (London: Longman, 1818), 431.

2. Alexander von Humboldt, *Reise durch Venezuela: Auswahl aus den amerikanischen Reisetagebüchern*, ed. Margot Faak (Berlin: Akademie, 2000), 81.

3. A. von Humboldt, *Reise durch Venezuela*, 81.

4. "Steatornis Caripensis," in Alexander von Humboldt, *Das graphische Gesamtwerk*, ed. Oliver Lubrich and Sarah Bärtschi, 3rd ed. (Darmstadt: Wissenschaftliche Buchgesellschaft, 2016), 243.

5. Pratt, *Imperial Eyes*, 123, 128, 132. For a thoughtful response, see Aaron Sachs, "The Ultimate 'Other': Post-Colonialism and Alexander von Humboldt's Ecological Relationship with Nature," *History and Theory* 42 (December 2003): 111–35. Londa Schiebinger offers a summary of recent research trends in "Forum Introduction: The European Colonial Science Complex," *Isis* 96 (2005): 52–55.

6. Franklin W. Knight, "Slavery in the Americas," in *A Companion to Latin American History*, ed. Thomas H. Holloway (Malden, MA: Wiley-Blackwell, 2011), 151.

7. Alexander von Humboldt, *Cuba-Werk*, ed. Hanno Beck, 2nd ed. (Darmstadt: Wissenschaftliche Buchgesellschaft, 2008), 156.

8. *Alexander von Humboldt. Lateinamerika am Vorabend der Unabhängigkeitsrevolution: Eine Anthologie von Impressionen und Urteilen aus seinen Reisetagebüchern*, ed. Margot Faak, 2nd ed. (Berlin: Akademie, 2003), 172–243.

9. John R. McNeill, *Mosquito Empires: Ecology and War in the Greater Caribbean, 1620–1914* (New York: Cambridge University Press, 2010), 3.

10. A. von Humboldt, *Reise durch Venezuela*, 330.

11. See the editions by Margot Faak and now Humboldt digital: Travel Journals, at https://edition-humboldt.de/reisetagebuecher/index.xql?&l=en.

12. See Jorge Cañizares-Esguerra, "How Derivative Was Humboldt? Microcosmic Nature Narratives in Early Modern Spanish America and the (Other) Origins of Humboldt's Ecological Sensibilities," in *Colonial Botany: Science, Commerce and Politics in the Early Modern World*, ed. Londa Schiebinger and Claudia Swan (Philadelphia: University of Pennsylvania Press, 2005),

148–65; Gregory T. Cushman, "Humboldtian Science, Creole Meteorology, and the Discovery of Human-Caused Climate Change in South America," *Osiris* 26 (2011): 19–44; Mark Thuner and Jorge Cañizares-Esguerra, eds., *The Invention of Humboldt: On the Geopolitics of Knowledge* (New York: Routledge, 2023). More cautious are Sachs, "The Ultimate 'Other,'" and Karl S. Zimmerer, "Mapping Mountains," in *Mapping Latin America: A Cartographic Reader*, ed. Jordana Dym and Karl Offen (Chicago: University of Chicago Press, 2011), 125–30.

13. A. von Humboldt, *Ueber die unterirdischen Gasarten*, 67, 201.

14. Alexander von Humboldt, *Amerikanische Reise*, ed. Hanno Beck, 6th ed. (Wiesbaden: Erdmann, 2009), 243, based on the estimate by Loren A. McIntyre.

15. Alexander von Humboldt, *Reise auf dem Río Magdalena, durch die Anden und Mexico*, vol. 2: *Übersetzung, Anmerkungen und Register*, trans. and ed. Margot Faak, 2nd ed. (Berlin: Akademie, 2003), 75.

16. A. von Humboldt, *Reise auf dem Río Magdalena*, 63.

17. Daniel Kehlmann, *Measuring the World*, trans. Carol Brown Janeway (New York: Pantheon Books, 2006).

18. A. von Humboldt, *Reise auf dem Río Magdalena*, 162.

19. Al. von Humboldt, *Reise in die Äquinoktial-Gegenden*, 2:1249–55. Cf. A. von Humboldt and Bonpland, *Personal Narrative of Travels to the Equinoctial Regions of the New Continent, during the Years 1799–1804*, trans. Helen Maria Williams, vol. 5, pt. 2 (London: Longman, 1821), 617–23.

20. A. von Humboldt to Johann Carl Freiesleben, August 1, 1804, in *Das Gute und Grosse wollen: Alexander von Humboldts amerikanische Briefe*, ed. Ulrike Moheit (Berlin: Rohrwall, 1999), 229.

Chapter 5: Delivering to the Public

1. Caroline von Humboldt to Wilhelm von Humboldt, August 28, 1804, in *Wilhelm und Caroline von Humboldt in ihren Briefen*, ed. Anna von Sydow, vol. 2: *Von der Vermählung bis zu Humboldts Scheiden aus Rom 1791–1808* (Berlin: Mittler, 1907), 231.

2. Alexander von Humboldt and Aimé Bonpland, *Essai sur la géographie des plantes; accompagné d'un tableau physique des régions équinoxiales, fondé sur des mesures exécutées, depuis le dixième degré la latitude boréale juasqu'au dixième degré de latitude australe, pendant les années 1799, 1800, 1801, 1802 et 1803* (Paris: Levrault,

Schoell, 1805), v. Cf. Alexander von Humboldt and Aimé Bonpland, *Essay on the Geography of Plants*, trans. Sylvie Romanowski, edited with an introduction by Stephen T. Jackson (Chicago: University of Chicago Press, 2009), 61.

3. A. von Humboldt to J. F. von Cotta, December 23, 1805, in *Alexander von Humboldt und Cotta*, 72, 71.

4. Alexander von Humboldt and Aimé Bonpland, *Ideen zu einer Geographie der Pflanzen nebst einem Naturgemälde der Tropenländer* (Tübingen: Cotta, 1807), 56. Cf. A. von Humboldt and Bonpland, *Essay on the Geography of Plants*, 86.

5. *Geschichte Berlins*. Vol. 2: *Von der Märzrevolution bis zur Gegenwart*, ed. Wolfgang Ribbe, 3rd ed. (Berlin: Berliner Wissenschafts-Verlag, 2002), 413.

6. A. von Humboldt to J. F. von Cotta, June 6, 1807, in *Alexander von Humboldt und Cotta*, 81.

7. A. von Humboldt and Bonpland, *Ideen zu einer Geographie der Pflanzen*, 2, 3–4. Cf. A. von Humboldt and Bonpland, *Essay on the Geography of Plants*, 64–65.

8. A. von Humboldt and Bonpland, *Ideen zu einer Geographie der Pflanzen*, 39. Cf. A. von Humboldt and Bonpland, *Essay on the Geography of Plants*, 79.

9. Alexander von Humboldt, *Ideen zu einer Physiognomik der Gewächse* (Tübingen: Cotta, 1806), 11.

10. Alexander von Humboldt, *Ansichten der Natur mit wissenschaftlichen Erläuterungen*, vol. 1 (Tübingen: Cotta, 1808), 172, 204. Cf. A. von Humboldt, *Views of Nature*, 159, 169.

11. A. von Humboldt und Bonpland, *Ideen zu einer Geographie der Pflanzen*, 44. Cf. A. von Humboldt and Bonpland, *Essay on the Geography of Plants*, 81.

12. Susanne S. Renner, Ulrich Päßler, and Pierre Moret, "'My Reputation Is at Stake': Humboldt's Mountain Plant Geography in the Making (1803–1825)," *Journal of the History of Biology* 56 (2023): 97–124. Cf. Zimmerer, "Mapping Mountains."

13. A. von Humboldt, *Ansichten der Natur* (1808), viii, referencing Friedrich Schiller, "Die Braut von Messina"; see *Friedrich Schiller: Sämtliche Werke*, vol. 2: *Dramen 2*, ed. Peter-André Alt, 3rd ed. (Munich: Hanser, 2013), 904. Cf. A. von Humboldt, *Views of Nature*, 26, and *The Dramas of Frederick Schiller*, trans. R. D. Boylan, Joseph Mellish, Anna Swanwick, and A. Lodge (London: Bell & Sons, 1920), 509.

14. A. von Humboldt, *Ansichten der Natur* (1808), 38. Cf. A. von Humboldt, *Views of Nature*, 40.

15. A. von Humboldt, *Ansichten der Natur* (1808), v; A. von Humboldt, *Kosmos: Entwurf einer physischen Weltbeschreibung*, vol. 1 (Stuttgart: Cotta, 1845), v. Cf. A. von Humboldt, *Views of Nature*, 25, and Alexander von Humboldt, *Cosmos: A Sketch of the Physical Description of the Universe*, trans. E. C. Otté (New York: Harper & Brothers, 1849), vol. 1, reprint with an introduction by Nicolaas A. Rupke (Baltimore: Johns Hopkins University Press, 1997), 7.

16. A. von Humboldt to Frederick William IV, March 23, 1841, in *Alexander von Humboldt – Friedrich Wilhelm IV.: Briefwechsel*, ed. Ulrike Leitner (Berlin: Akademie, 2013), 202.

17. Kurt-R. Biermann, *Alexander von Humboldt*, 4th, rev. ed. (Berlin: Teubner, 1990), 91.

18. A. von Humboldt, *Ansichten der Natur* (1808), v. Cf. A. von Humboldt, *Views of Nature*, 25.

19. Cf. Horst Fiedler and Ulrike Leitner, eds., *Alexander von Humboldts Schriften: Bibliographie der selbständig erschienenen Werke* (Berlin: Akademie, 2000), 69.

20. Alexander von Humboldt, *Versuch über den politischen Zustand des Königreichs Neu-Spanien . . .* , vol. 1 (Tübingen: Cotta, 1809), 53. Cf. Alexander von Humboldt, *Political Essay on the Kingdom of New Spain: A Critical Edition*, edited with an introduction by Vera M. Kutzinski and Ottmar Ette, trans. J. Ryan Poynter, Kenneth Berri, and Vera M. Kutzinski, vol. 1 (Chicago: University of Chicago Press, 2019), 183 (translated here as "tableau").

21. A. von Humboldt, *Versuch über den politischen Zustand des Königreichs Neu-Spanien*, 1:106. Cf. A. von Humboldt, *Political Essay on the Kingdom of New Spain*, 1:230.

22. A. von Humboldt to Johann Wolfgang Goethe, January 3, 1810, quoted in Fiedler and Leitner, *Alexander von Humboldts Schriften*, 134.

23. A. von Humboldt, *Ideen zu einer Geographie der Pflanzen*, iv. This passage is not included in A. von Humboldt and Bonpland, *Essay on the Geography of Plants*, 61.

24. Hanno Beck, *Alexander von Humboldt*, vol. 2: *Vom Reisewerk zum "Kosmos" 1804–1859* (Wiesbaden: Steiner, 1961), 59.

25. Alexander von Humboldt and Jean-Baptiste Biot, "Sur les variations du magnétism terrestre à différentes latitudes" [1804], in Alexander von

Humboldt, *Sämtliche Schriften*, vol. 3: *1810–1819*, ed. Michael Strobl and Jobst Welge (Munich: dtv, 2019), 276–97. See Alexander von Humboldt, *Schriften zur physikalischen Geographie*, ed. Hanno Beck, 2nd ed. (Darmstadt: Wissenschaftliche Buchgesellschaft, 2008), 196–97, 200–201.

26. Alexander von Humboldt, "Des lignes isothermes et de la distribution de la chaleur sur le globe" [1817], in A. von Humboldt, *Sämtliche Schriften*, vol 3: *1810–1819*, 471–533.

27. A. von Humboldt, *Versuch über den politischen Zustand des Königreichs Neu-Spanien*, 1:xvi. Cf. A. von Humboldt, *Political Essay on the Kingdom of New Spain*, 1:22.

28. Alexander von Humboldt, *Political Essay on the Island of Cuba*, ed. Vera M. Kutzinski and Ottmar Ette (Chicago: University of Chicago Press, 2011), 81.

29. Jürgen Osterhammel, *Unfabling the East: The Enlightenment's Encounter with Asia*, trans. Robert Savage (Princeton: Princeton University Press, 2018), 489.

Chapter 6: I Love What I Comprehend

1. Reinhard Stauber, *Der Wiener Kongress* (Vienna: Böhlau, 2014), 181.

2. Biermann, *Alexander von Humboldt*, 69–70.

3. *Geschichte Berlins*, vol. 1: *Von der Frühgeschichte bis zur Industrialisierung*, ed. Wolfgang Ribbe, 3rd ed. (Berlin: Berliner Wissenschafts-Verlag, 2002), 480.

4. *Geschichte Berlins*, 1:532.

5. Andreas W. Daum, "Resonance Space: The 'Cosmos Lectures,'" in *Wilhelm and Alexander von Humboldt: Berlin Cosmos*, ed. Paul Spies, Ute Tintemann, and Jan Mende, trans. Gérard Goodrow (Cologne: Wiegand, 2020), 148–51. For a new edition of the transcripts, see Alexander von Humboldt and Henriette Kohlrausch, *Die Kosmos-Vorlesung an der Berliner Sing-Akademie*, ed. Christian Kassung and Christian Thomas (Berlin: Insel, 2019).

6. A. von Humboldt to J. F. von Cotta, March 1, 1828, in *Alexander von Humboldt und Cotta*, 160.

7. A. von Humboldt, and Kohlrausch, *Kosmos-Vorlesung*, 283.

8. So the subtitle of his *Cosmos*.

9. A. von Humboldt to Karl August Varnhagen von Ense, October 24, 1834, in *Letters of Alexander von Humboldt to Varnhagen von Ense: From 1827 to 1858*, trans. Friedrich Kapp (New York: Rudd & Carleton, 1860), 35–36.

10. See Andreas W. Daum, *Wissenschaftspopularisierung im 19. Jahrhundert: Bürgerliche Kultur, naturwissenschaftliche Bildung und die deutsche Öffentlichkeit, 1848–1914* (1998; Munich: Oldenbourg, 2002).

11. Daum, *Wissenschaftspopularisierung*, 89–103.

12. Andreas W. Daum, "Varieties of Popular Science and the Transformations of Public Knowledge: Some Historical Reflections," *Isis* 100 (June 2009): 319–32.

13. A. von Humboldt to Wilhelm von Humboldt, June 9/21, 1829, in *Alexander von Humboldt: Briefe aus Russland 1829*, ed. Eberhard Knobloch, Ingo Schwarz, and Christian Suckow (Berlin: Akademie, 2009), 138.

14. "Rede, gehalten von Herrn Alexander von Humboldt in der außerordentlichen Sitzung der Kaiserlichen Akademie der Wissenschaften zu Sankt Petersburg am 16./28.11. 1829," in *Alexander von Humboldt: Briefe aus Russland*, 266–85.

15. Cannon, "Humboldtian Science." Cf. Daum, "Humboldtian Science and Humboldt's Science."

16. Carl Friedrich Zelter to Johann Wolfgang von Goethe, February 2, 1830, in *Briefwechsel zwischen Goethe und Zelter in den Jahren 1799 bis 1832*, Goethe: Münchener Ausgabe, vol. 20.2, ed. Hans-Günter Ottenberg and Edith Zehm (Munich: Hanser, 1991), 1316.

17. *Alexander von Humboldts Reise durchs Baltikum nach Russland und Sibirien 1829*, ed. Hanno Beck, 6th ed. (Wiesbaden: Erdmann, 2009), 163.

18. A. v. Humboldt's *Fragmente einer Geologie und Klimatologie Asiens: Aus dem Französischen mit Anmerkungen, einer Karte und einer Tabelle vermehrt von Julius Loewenberg* (Berlin: List, 1832), 17, 7.

19. A. v. Humboldt's *Fragmente einer Geologie und Klimatologie Asiens*, 180.

20. A. von Humboldt to J. Schulze, early 1843, quoted in Fiedler and Leitner, *Alexander von Humboldts Schriften*, 355.

21. A. von Humboldt, *Zentral-Asien: Untersuchungen zu den Gebirgsketten und zur vergleichenden Klimatologie. Nach der Übersetzung Wilhelm Mahlmanns aus dem Jahr 1844*, ed. Oliver Lubrich (Frankfurt a.M.: S. Fischer, 2009), 7.

22. Laura Péaud, "Die diplomatischen Berichte Alexander von Humboldts aus Paris zwischen 1835 und 1847," in *"Mein zweites Vaterland": Alexander von Humboldt und Frankreich*, ed. David Blankenstein et al. (Berlin: de Gruyter, 2015), 15–31.

23. David Blankenstein et al., "Vorwort," in *"Mein zweites Vaterland,"* viii.

24. A. von Humboldt to August Böckh, July 24, 1837, in *Alexander von Humboldt – August Böckh: Briefwechsel*, ed. Romy Werther (Berlin: Akademie, 2011), 88.

25. Alexander von Humboldt, *Kritische Untersuchungen über die historische Entwickelung der geographischen Kenntnisse von der Neuen Welt und die Fortschritte der nautischen Astronomie in dem 15ten und 16ten Jahrhundert von Alexander von Humboldt*, vol. 1 (Berlin: Nicolai, 1836), 8.

26. A. von Humboldt to Karl August Varnhagen von Ense, October 22, 1837, in *Letters of Alexander von Humboldt*, 61; A. von Humboldt to J. G. von Cotta, February 28, 1838, in *Alexander von Humboldt und Cotta*, 204.

27. A. von Humboldt to J. G. von Cotta, June 29, 1838, in *Alexander von Humboldt und Cotta*, 207.

28. A. von Humboldt to J. G. von Cotta, May 5, 1839, in *Alexander von Humboldt und Cotta*, 212.

Chapter 7: Between Cosmos and Fragments

1. Janet Browne, *Charles Darwin: Voyaging* (Princeton: Princeton University Press, 1995), 362–63, 436–39.

2. Alexander von Humboldt, *Kosmos*, vol. 1, 1845, v. Cf. A. von Humboldt, *Cosmos*, vol. 1 (1997), 7.

3. A. von Humboldt, *Kosmos*, vol. 1, vi. Cf. A. von Humboldt, *Cosmos*, vol. 1, 7.

4. A. von Humboldt, *Kosmos*, vol. 1, 79. Cf. A. von Humboldt, *Cosmos*, vol. 1, 79.

5. A. von Humboldt, *Kosmos*, vol. 1, viii. Cf. A. von Humboldt, *Cosmos*, vol. 1, 8.

6. A. von Humboldt, *Kosmos*, vol. 1, viii, 3. Cf. A. von Humboldt, *Cosmos*, vol. 1, 9, 23.

7. Daum, *Wissenschaftspopularisierung*, 277.

8. *Briefe über Alexander von Humboldt's Kosmos: Ein Commentar zu diesem Werke für gebildete Laien*, vol. 1, ed. Bernhard Cotta, 2nd rev. ed. (Leipzig: Weigel, 1850), vii.

9. Hermann Klencke, *Mikroskopische Bilder: Naturansichten aus dem kleinsten Raume. Ein Gemälde des Mikrokosmos in seinen Gestalten und Gesetzen in Briefen an Gebildete* (Leipzig: Weber, 1853), xi; cf. ix.

10. Daum, *Wissenschaftspopularisierung*, 280–86, 460.

11. Daum, *Wissenschaftspopularisierung*, 138–67; Andreas W. Daum, "Science, Politics, and Religion: Humboldtian Thinking and the Transformations of Civil Society in Germany, 1830–1870," in *Science and Civil Society*, ed. Tom Broman and Lynn Nyhart (*Osiris*, no. 17), 107–40 (Chicago: University of Chicago Press, 2002).

12. A. von Humboldt to Gabriele von Bülow, March 22, 1848, in *Alexander von Humboldt – Gabriele von Bülow: Briefe*, ed. Ulrike Leitner (Berlin: de Gruyter, 2023), 257.

13. A. von Humboldt to August von Hedemann, September 18, 1848, German Literature Archive, Marbach, Manuscripts Division, Nr. 62.2140.

14. *Geschichte Berlins*, 2:657.

15. A. von Humboldt to J. G. von Cotta, November 20, 1854, in *Alexander von Humboldt und Cotta*, 551.

16. A. von Humboldt to J. G. von Cotta, October 29, 1853, in *Alexander von Humboldt und Cotta*, 517.

17. Already in Alexander von Humboldt, *Kosmos: Entwurf einer physischen Weltbeschreibung*, vol. 2 (Stuttgart: Cotta, 1847), 27, 144, 252. Cf. Alexander von Humboldt, *Cosmos: A Sketch of the Physical Description of the Universe*, trans. E. C. Otté, vol. 2 (New York: Harper & Brothers, 1850), reprinted with an introduction by Michael Dettelbach (Baltimore: Johns Hopkins University Press, 1997), 39, 113, 212.

18. Petra Werner, *Himmel und Erde: Alexander von Humboldt und sein Kosmos* (Berlin: Akademie, 2004), 292; Petra Werner, *Gasanova ohne Frauen? Bemerkungen zu Alexander von Humboldts Korrespondenzpartnerinnen* (Berlin: Alexander-von-Humboldt-Forschungstelle, 2000).

19. A. von Humboldt to Gabriele von Bülow, December 26, 1847, in *Alexander von Humboldt – Gabriele von Bülow*, 237; A. von Humboldt to his grandnephew Wilhelm von Humboldt, March 4, 1857, German Literature Archives, Marbach, Manuscripts Division, Nr. 2005.33.1.

20. "Ruf um Hülfe," *Berlinische Nachrichten von Staats- und gelehrten Sachen*, March 20, 1859.

21. Rupke, *Alexander von Humboldt*.

22. Daum, *Wissenschaftspopularisierung*, 142–52, 203–7.

23. Andreas W. Daum, "Nation, Naturforschung und Monument: Humboldt-Denkmäler in Deutschland und den USA," in *Die Kunst der Geschichte: Historiographie, Ästhetik, Erzählung*, ed. Martin Baumeister et al. (Göttingen: Vandenhoeck & Ruprecht, 2009), 99–124.

24. A. von Humboldt to J. G. von Cotta, October 31, 1854, in *Alexander von Humboldt und Cotta*, 545.

25. Lorraine Daston, "The Humboldtian Gaze," in *Science as Cultural Practice*, vol. 1: *Cultures and Politics of Research from the Early Modern Period to the Age of Extremes*, ed. Moritz Epple and Claus Zittel (Berlin: De Gruyter, 2010), 45.

26. Charles P. Snow, *The Two Cultures*, with an introduction by Stefan Collini (New York: Cambridge University Press, 2012).

SELECTED BIBLIOGRAPHY

Books by Alexander von Humboldt

An indispensable tool for identifying and understanding Humboldt's book publications is *Alexander von Humboldts Schriften: Bibliographie der selbständig erschienenen Werke*, a comprehensive bibliography edited by Horst Fiedler und Ulrike Leitner (Berlin: Akademie, 2000). The list below provides an overview and follows the chronology of the publication dates. Recent English-language translations, including abridgments, are mentioned below selected titles. None of Humboldt's books from the years 1790 to 1799 has yet been translated into English.

Mineralogische Beobachtungen über einige Basalte am Rhein (1790)
Florae Fribergensis Specimen Plantas Cryptogamicas Praesertim Subterraneas Exhibens (1793)
Aphorismen aus der chemischen Physiologie der Pflanzen (1794)
Versuche über die gereizte Muskel- und Nervenfaser nebst Vermuthungen über den chemischen Process des Lebens in der Thier- und Pflanzenwelt, 2 vols. (1797–99)
Ueber die unterirdischen Gasarten und die Mittel, ihren Nachtheil zu vermindern: Ein Beytrag zur Physik der praktischen Bergbaukunde (1799)
Versuche über die chemische Zerlegung des Luftkreises und über einige andere Gegenstände der Naturlehre (1799)
Essai sur la géographie des plantes (1805–7); first German edition as *Ideen zu einer Geographie der Pflanzen* (1807)
   Essay on the Geography of Plants (University of Chicago Press, 2009)
Ansichten der Natur, mit wissenschaftlichen Erläuterungen (1808; second and expanded ed. 1826; third and expanded ed. 1849)
   Views of Nature (University of Chicago Press, 2014)

Essai politique sur le Royaume de la Nouvelle-Espagne (1811)

   Political Essay on the Kingdom of New Spain, 2 vols. (University of Chicago Press, 2019)

Relations historique (1814–31), part of *Voyage aux régions équinoxiales du nouveau continent*

   Personal Narrative of Travels to the Equinoctial Regions of the New Continent; trans. Helen Maria Williams (7 vols., Longman 1814–29)

   Personal Narrative of Travels to the Equinoctial Regions of America; trans. Thomasina Ross (abridged, 3 vols., Bohn 1852–53)

   Personal Narrative of a Journey to the Equinoctial Regions of the New Continent; trans. Jason Wilson (abridged, Penguin 1995)

Atlas géographique et physique des régions équinoxiales du nouveau continent (1814–38)

Vues des Cordillères et monuments des peuples indigènes de l'Amérique (1816); partial German edition as *Pittoreske Ansichten der Cordilleren* (1810)

   Views of the Cordilleras and Monuments of the Indigenous Peoples of the Americas (University of Chicago Press, 2011)

Essai géognostique sur le gisement des roches dans les deux hémisphères (1823)

   A Geognostical Essay on the Superposition of Rocks in Both Hemispheres (1823)

Essai politique sur l'île de Cuba (1826)

   Political Essay on the Island of Cuba (University of Chicago Press, 2011)

Fragmens de géologie et de climatologie asiatiques (1831); in German as *Fragmente einer Geologie und Klimatologie Asiens* (1832)

Examen critique de l'histoire de la géographie du nouveau continent (1834–38); in German as *Kritische Untersuchungen über die historische Entwickelung der geographischen Kenntnisse von der Neuen Welt* (1836–52)

Asie centrale, 3 vols. (1843)

Kosmos: Entwurf einer physischen Weltbeschreibung, 5 vols. (1845–62)

   Cosmos: Sketch of a Physical Description of the Universe; trans. Edward Sabine, 4 vols. (Longman, 1846–1858)

   Cosmos: A Sketch of the Physical Description of the Universe; trans. E. C. Otté, 5 vols. (Bohn 1848–58; reprint of vols. 1–2, Johns Hopkins University Press, 1997)

Kleinere Schriften (1853)

Selected Secondary Literature

Beck, Hanno. *Alexander von Humboldt*. 2 vols. Wiesbaden: Steiner, 1959–61.

Blackbourn, David. *Germany in the World: A Global History, 1500–2000*. New York: Liveright, 2023.

———. *History of Germany, 1780–1918: The Long Nineteenth Century*. 2nd ed. Malden, MA: Blackwell, 2003.

Blankenstein, David, et al., eds. *"Mein Zweites Vaterland": Alexander von Humboldt und Frankreich*. Berlin: de Gruyter, 2015.

Bleichmar, Daniela. *Visible Empire: Botanical Expeditions and Visual Culture in the Hispanic Enlightenment*. Chicago: University of Chicago Press, 2012.

Browne, Janet. *Charles Darwin*. Vol. 1: *Voyaging*. Vol. 2: *The Power of Place*. Princeton: Princeton University Press, 1995–2002.

Cañizares-Esguerra, Jorge. "How Derivative Was Humboldt? Microcosmic Nature Narratives in Early Modern Spanish America and the (Other) Origins of Humboldt's Ecological Sensibilities." In *Colonial Botany: Science, Commerce and Politics in the Early Modern World*, ed. Londa Schiebinger and Claudia Swan, 148–65. Philadelphia: University of Pennsylvania Press, 2005.

———. *Nature, Empire, and Nation: Explorations of the History of Science in the Iberian World*. Stanford: Stanford University Press, 2006.

Cannon, Susan Faye. "Humboldtian Science." In *Science in Culture: The Early Victorian Period*, 73–110. New York: Science History Publications, 1978.

Clark, Christopher. *Iron Kingdom: The Rise and Downfall of Prussia, 1600–1947*. Cambridge, MA: Belknap Press, 2006.

Clark, William, Jan Golinski, and Simon Schaffer, eds. *The Sciences in Enlightened Europe*. Chicago: University of Chicago Press, 1999.

Daston, Lorraine. "The Humboldtian Gaze." In *Science as Cultural Practice*. Vol. 1: *Cultures and Politics of Research from the Early Modern Period to the Age of Extremes*, ed. Moritz Epple and Claus Zittel, 45–66. Berlin: De Gruyter, 2010.

Daum, Andreas W. "German Naturalists in the Pacific around 1800: Entanglement, Autonomy, and a Transnational Culture of Expertise." In *Explorations and Entanglements: Germans in Pacific Worlds from the Early Modern Period to World War I*, ed. Hartmut Berghoff, Frank Biess, and Ulrike Strasser, 79–102. New York: Berghahn Books, 2019.

———. "Humboldtian Science and Humboldt's Science." *History of Science* 62 (2024).

———. "Science, Politics, and Religion: Humboldtian Thinking and the Transformations of Civil Society in Germany, 1830–1870." In *Science and Civil Society*, ed. Tom Broman and Lynn Nyhart (*Osiris*, no. 17), 138–67. Chicago: University of Chicago Press, 2002.

———. "Social Relations, Shared Practices, and Emotions: Alexander von Humboldt's Excursion into Literary Classicism and the Challenges to Science around 1800." *Journal of Modern History* 91 (March 2019): 1–37.

———. "Varieties of Popular Science and the Transformations of Public Knowledge: Some Historical Reflections." *Isis* 100 (June 2009): 319–32.

———. "Wissenschaft and Knowledge." In *Germany, 1800–1870*, ed. Jonathan Sperber, 137–61. Oxford: Oxford University Press, 2004.

———. *Wissenschaftspopularisierung im 19. Jahrhundert: Bürgerliche Kultur, naturwissenschaftliche Bildung und die deutsche Öffentlichkeit, 1848–1914.* 1998. 2nd ed. Munich: Oldenbourg, 2002.

Dettelbach, Michael. "Humboldtian Science." In *Cultures of Natural History*, ed. N. Jardine, J. Secord, and E. C. Sparry, 287–304. Cambridge: Cambridge University Press, 1996.

Drayton, Richard. *Nature's Government: Science, Imperial Britain and the "Improvement" of the World.* New Haven: Yale University Press, 2000.

Glaubrecht, Matthias. "'Through a Country We Never Intended to See': Revisiting the Humboldt Renaissance." In *Alexander von Humboldt: Multiperspective Approaches*, ed. Gregor C. Falk, Manfred R. Strecker, and Simon Schneider, 3–36. Cham: Springer, 2022.

Goldstein, Jürgen. *Georg Forster: Voyager, Naturalist, Revolutionary.* Chicago: University of Chicago Press, 2019.

Home, R. W. "Humboldtian Science Revisited: An Australian Case Study." *History of Science* 33 (March 1995): 1–22.

Klein, Ursula. *Humboldts Preußen: Wissenschaft und Technik im Aufbruch.* Darmstadt: Wissenschaftliche Buchgesellschaft, 2015.

———. "The Prussian Mining Official Alexander von Humboldt." *Annals of Science* 69 (2012): 27–68.

Kontje, Todd. *Georg Forster: German Cosmopolitan.* University Park: Pennsylvania State University Press, 2022.

Lack, H. Walter. *Alexander von Humboldt: The Botanical Exploration of the Americas.* Trans. Stephen Telfer. Munich: Prestel, 2018.

Liebersohn, Harry. *The Traveler's World: Europe to the Pacific*. Cambridge, MA: Harvard University Press, 2006.

Manning, Patrick, and Daniel Rood, eds. *Global Scientific Practice in an Age of Revolutions, 1750–1850*. Pittsburgh: University of Pittsburgh Press, 2016.

Marchand, Suzanne L. *Porcelain: A History from the Heart of Europe*. Princeton: Princeton University Press, 2020.

Martin, Alison E. *Nature Translated: Alexander von Humboldt's Works in Nineteenth-Century Britain*. Edinburgh: Edinburgh University Press, 2018.

McNeill, John R. *Mosquito Empires: Ecology and War in the Greater Caribbean, 1620–1914*. New York: Cambridge University Press, 2010.

Nipperdey, Thomas. *Germany from Napoleon to Bismarck: 1800–1866*. Trans. Daniel Nolan. Princeton: Princeton University Press, 1996.

Olesko, Kathryn. "Germany." In *The Cambridge History of Science*. Vol. 8: *Modern Science in National, Transnational, and Global Contexts*, ed. Hugh Richard Slotten, Ronald L. Numbers, and David N. Livingstone, 233–77. Cambridge: Cambridge University Press, 2020.

———. "Humboldtian Science." In *The Oxford Guide to the History of Physics and Astronomy*, ed. John L. Heilbron, 159–62. New York: Oxford University Press, 2005.

Osterhammel, Jürgen. *The Transformation of the World: A Global History of the Nineteenth Century*. Trans. P. Camiller. Princeton: Princeton University Press, 2014.

———. *Unfabling the East: The Enlightenment's Encounter with Asia*. Trans. Robert Savage. Princeton: Princeton University Press, 2018.

Outram, Dorinda. *The Enlightenment*. 4th ed. New York: Cambridge University Press, 2019.

Pratt, Mary Louise. *Imperial Eyes: Travel Writing and Transculturation*. 2nd ed. London: Routledge, 2009.

Reill, Peter Hanns. *Vitalizing Nature in the Enlightenment*. Berkeley: University of California Press, 2005.

Renner, Susanne S., Ulrich Päßler, and Pierre Moret. "'My Reputation Is at Stake': Humboldt's Mountain Plant Geography in the Making (1803–1825)." *Journal of the History of Biology* 56 (2023): 97–124.

Ribbe, Wolfgang, ed. *Geschichte Berlins*. 2 vols. 3rd ed. Berlin: Berliner Wissenschafts-Verlag, 2002.

Richards, Robert J. *The Romantic Conception of Life: Science and Philosophy in the Age of Goethe*. Chicago: University of Chicago Press, 2002.

Rupke, Nicolaas A. *Alexander von Humboldt: A Metabiography*. Chicago: University of Chicago Press, 2008.

———. "Humboldtian Distribution Maps: The Spatial Ordering of Scientific Knowledge." In *The Structure of Knowledge: Classifications of Science and Learning since the Renaissance*, ed. Tore Frängsmyr Tore, 93–116. Berkeley: Office for History of Science and Technology, 2001.

———. "Humboldtian Medicine." *Medical History* 40 (1996): 293–310.

Sachs, Aaron. "The Ultimate 'Other': Post-Colonialism and Alexander von Humboldt's Ecological Relationship with Nature." *History and Theory* 42 (December 2003): 111–35.

Safier, Neil. *Measuring the New World: Enlightenment Science and South America*. Chicago: University of Chicago Press, 2008.

Schiebinger, Londa. "Forum Introduction: The European Colonial Science Complex." *Isis* 96 (2005): 52–55.

Schwarz, Ingo. "From Alexander von Humboldt's Correspondence with Thomas Jefferson and Albert Gallatin." *Berliner Manuskripte zur Alexander von Humboldt Forschung* 2 (2000): 1–20.

Smith, Helmut Walser, *Germany: A Nation in Its Time. Before, During, and After Nationalism, 1500–2000*. New York: Norton, 2020.

Stauber, Reinhard. *Der Wiener Kongress*. Vienna: Böhlau, 2014.

Thuner, Mark, and Jorge Cañizares-Esguerra, eds. *The Invention of Humboldt: On the Geopolitics of Knowledge*. New York: Routledge, 2023.

Wulf, Andrea. *The Invention of Nature: Alexander von Humboldt's New World*. New York: Knopf, 2015.

Zantop, Susanne. *Colonial Fantasies: Conquest, Family, and Nation in Precolonial Germany, 1770–1870*. Durham: Duke University Press, 1997.

Zeuske, Michael. "Humboldt in Venezuela and Cuba: The 'Second Slavery.'" *German Life and Letters* 74 (July 2021): 311–25.

Zimmerer, Karl S. "Mapping Mountains." In *Mapping Latin America: A Cartographic Reader*, ed. Jordana Dym and Karl Offen, 125–30. Chicago: University of Chicago Press, 2011.

ILLUSTRATIONS

Figures

Maps

INDEX

Page numbers in *italics* indicate figures and tables.